SHEARING LIFE *in* AUSTRALIA

SHEEP GRAZING ON LOVELY, ROLLING COUNTRYSIDE.

SHEARING LIFE *in* AUSTRALIA

RAY SHERMAN

Kangaroo Press

Designed by Linda Robertshaw

First published in 1997 by Kangaroo Press
an imprint of Simon & Schuster Australia
20 Barcoo Street East Roseville NSW 2069
Printed in Hong Kong through Colorcraft Ltd

ISBN 0 86417 874 3

Contents

THE RAINBOW INFORMS US THAT IT IS ALREADY RAINING.

Acknowledgments

I wish to acknowledge the assistance given to me by Ivan Letchford in the preparation of this book. He was not just a fund of information. He was patient, hospitable and always available. I would be remiss if I did not thank Ivan's wife, Madge, who regularly manoeuvred my visits to coincide with meals designed to sate a gourmand.

I would like to acknowledge the worth of all the material read in preparing Shearing Life in Australia. When I began this project with Ivan Letchford and his shearers in mind I wanted to increase my knowledge of the wool industry and to do so in a brisk fashion. Not only did I learn about the development and progress of the wool industry, I learned a great deal about the history of Australia. There is no doubt that sheep and wool have been intimately associated with the growth and maturity, the culture and the economy of this 'wide brown land'. To all those who have written before me: thank you. I am grateful indeed.

Song and verse were taken from *Favourite Australian Bush Songs* by Lionel Long and Graham Jenkin, *A Treasury of Favourite Australian Songs* by Therese Radic and *The Overlander Song Book* by Ron Edwards.

Family Circle magazine kindly allowed me to reprint my article 'The Shearers' Cook'. The Macmillan Company of Australia Pty Ltd allowed me to quote from Keith Willey's *The Drovers* upon payment of a moderate fee.

Veronica Nass rendered yeoman service in coaxing a word processor through its paces.

I used Nikon cameras and lenses, the equipment earning an A-plus for consistency and reliability. Colour reversal films were Fuji or Kodak, encompassing speeds of 64 to 400.

Black and white historical photographs were kindly provided by the National Library of Australia.

To the members of shearing teams who spoke freely and did not complain when I was in their way, thank you. I trust that I have created a true representation of your lives.

To the memory of Keith Willey

A talented writer and storyteller

A true blue Australian

Not least, a fine friend

If I had been a poet instead of a worker with the brush, I should have described the scattered flocks on sunlit plains and gum-covered ranges, the coming of spring, the gradual massing of the sheep towards that one centre, the woolshed, through which the accumulated growth and wealth of the year is carried; the shouts of the men, the galloping of horses and the barking of dogs as the thousands are driven, half seen, through the hot dust cloud, to the yards; then the final act, and the dispersion of the denuded sheep.

Tom Roberts, circa 1890, referring
to his painting, *Shearing the Rams.*

The friends are too warm, the whisky too strong, and the seats too soft for Tommy Ryan. His place is out among the shearers of the billabongs.

Thomas Ryan, a shearer and an
Australian Workers' Union (AWU)
official, upon his retirement from
the Queensland Parliament in 1893.

Introduction

Sir David Attenborough observes that man began worshipping animals and ended by enslaving them. As I heard Sir David speak these words I thought of how accurately he had outlined the wool industry in Australia. As a nation Australia certainly has worshipped the sheep. It is common knowledge that the country 'rode on the sheep's back' for much of its formative period. The prosperity which came to a distant English colony resulted from the widespread establishment of sheep grazing over the hinterlands of a vast island continent.

Many cycles of halcyon days mixed with economic as well as climatic disasters have followed the man on the land. Today, over 200 years after the first sheep stepped on to Antipodean soil, their descendants are enslaved within a system which depends upon their fleece—and to a lesser extent their very flesh—for a solid percentage of the nation's export earnings.

Often it seems that the principals in this Hollywood-sized megacast are but passive bit players. Australian historian C. E. W. Bean remarked that a sheep had about five minutes of singular importance per year: the time that he is being handled by a shearer. 'From the moment he is lambed a sheep is either so much wool or so much meat; and wool or meat he remains until he dies.' In a land where there are ten sheep for each person perhaps it cannot be any other way.

Centre stage in the wool theatre is held by people: originally by those settlers who were anything and everything from bullocky drivers to shearers to cooks to squatters and selectors; latterly by updated versions of the same toilers, for much of our drama is conducted today as it was in the early years.

Two traits have always distinguished the wool scene: hard work and a profusion of real characters. Technology may dominate many industries but this is not so where sheep are concerned.

A shearers' cook told me that the First Fleet put down anchors, unloaded their sheep and the men to look after them, and since that day everything has continued in a straight line, unchanged. The nature of this toil generates ravenous appetites and gargantuan thirsts. Our plot contains bush stories from over the past two centuries, stories which are part of Australia's history. America has its cowboys, Britain its gamekeepers, Canada its mounted policemen, but Australia has its shearers—people carrying on and creating fresh folklore which, even today, contributes to the Australian psyche. Our approach then is threefold: to look at the activities of today's sheepmen, to review the historical developments which trace the industry, and to be entertained by the larger-than-life figures who have become heroic in the Australian consciousness. We will be aided and abetted along our way by Ivan Letchford. Ivan is a redoubtable shearing contractor of many years, an erudite spokesman for his industry, a gentleman and a friend, not necessarily in that order. He has served the Shearing Contractors

PORTIONS OF AUSTRALIA WERE OVER-CLEARED FOR CULTIVATION PURPOSES BY THE ORIGINAL SETTLERS. THE RESULT WAS THAT THESE AREAS BECAME MARGINAL FARMING COUNTRY WHEN THE OWNERS DECIDED TO RUN SHEEP ON THE PROPERTIES.

Association of Australia as its Industrial Officer for the past sixteen years and has been its President for four years.

Men and women, then and now, the famous and the unknown, their animals, their exploits, their stories, are the core of this book about the wool industry in Australia over two centuries after the nation's founding.

1

The BEASTS *of the* FIELD

Historical Notes

The very word 'sheep' has a long history. The spelling was considerably different in the stem word, the Old German *scaf*. This became *scaep*, *seeap*, *scep* and other words in the Anglo-Saxon translation. Today, the German word for sheep is *schaf* while in Dutch it is a quite recognisable *schaep*.

Those early merchandisers, the Phoenicians, as well as being the first group to acquaint Europeans with tame sheep, were known to trade in woollen cloth. It has been written that a sizeable flock, perhaps numbering as many as 2000 sheep, grazed close to Memphis. That's Egypt, not Tennessee.

One might think Jason and the Argonauts are a snappily named rock group. In fact, Jason and his fifty-man crew in the oared *Argo* are part of Greek mythology. It is a convoluted and unhappy tale but here are bits from the later portion, reduced to a manageable size. A sheep covered with a remarkable Golden Fleece was slain in Colchis on the Black Sea coast. Jason was directed to bring back the Golden Fleece on board the *Argo*. This he did with the assistance of Medea, a sorceress who had fallen in love with him. Jason returned to Greece, Golden Fleece in hand, to receive a hero's welcome. Everything went swimmingly until Jason got his knickers in a knot over Glauce, the King's daughter. Medea wasn't too keen on this turn of events and so she put her magical powers to play. Glauce burned to death. Jason didn't fare any better. The incantatory Medea avenged his amorous transgression by causing part of the beached *Argo* to fall on him, crushing him to death. Trust the ancients to turn a little fun into a whacking great tragedy.

Sheep were said to be important in Middle East agriculture, and they featured in biblical stories. The Israelites apparently managed to plunder 675 000 sheep in a raid on Midian and in another campaign, this time against the Hagarites, they seized 250 000 sheep. The Old Testament's Jacob was an astute sheep breeder, so adept that he was able to selectively breed in order to enhance certain traits in his flocks.

George Brown, in his *Sheep Breeding in Australia*, likened ancient Hebrew life to Aussie bush life. The dry climate of Australia resembles that of the Near East. Water is a scarce commodity in both societies and drought a common concern.

During this time in history shearing was done annually as part of a gala occasion. Shears were probably invented in the Middle East around 1000 BC. Prior to that time, domestic sheep had been combed or plucked to harvest their wool. There was a transitional period when the process altered from combing to shearing. Scholar Leslie White found that in Greek the word *peko* means both 'to comb' and 'to shear', obviously language alluding to the blending of the old method with the new. White also discusses the Dead Sea Scrolls which were unearthed between 1947 and 1956 in the Near East. Dating from the

The Springtime it Brings on the Shearing

E.J. OVERBURY

The springtime it brings on the shearing,
And it's then you will see them in droves,
To the west-country stations all steering,
A-seeking a job off the coves.

Chorus:
With my raggedy old swag on my shoulder
And a billy quart-pot in my hand,
I tell you we'll 'stonish the new chums,
When they see how we travel the land.

From Boonabri up to the border,
Then it's over to Bourke; there and back.
On the hills and the plains you will see them,
The men on the Wallaby Track.

And after the shearing is over
And the wool season's all at an end,
It is then you will see the flash shearers
Making johnny-cakes round in the bend.

approximate time of Christ's life, the scrolls are actually fashioned from sheep and goat skins. Under microscopic examination of the parchments, wool fibres of varying nature could be detected. Of interest is the fact that some samples of these wool fibres are very much like those found on today's modem Merino.

Sheep

Sheep are ruminants; that is, animals with four stomachs. Sheep resemble cows in that they both chew the cud. It works like this: the sheep bites off a mouthful of grass, swallows it without chewing and later, when resting, regurgitates the grass (cud) which is then slowly chewed and swallowed again. After all this the grass returns to the first stomach, a large compartment called the rumen.

Inside the rumen bacteria attack and digest the fibre in the sheep's food. Chewing the cud makes it easier for the bacteria to break down the fibre because it reduces it to minute pieces. A lamb has no rumen at birth; it generally appears over the next several months. Since the young lamb can digest only milk, a special groove delivers it directly to the fourth stomach.

Another interesting aspect concerning the rumen is its ability to fine-tune the diet of the sheep. The bacteria living in the rumen are able to produce a number of amino acids as well as vitamin B. If we had this ability we could eat junk food and convert it into a healthy meal.

The second stomach is called the reticulum. Its task is to manage the amount of water in the rumen. The omasum or psalterium is the third chamber while the abomasum is the fourth and final stomach. Here the digested food is taken into the bloodstream. Young grass is high in cellulose, a type of fibre that sheep can digest readily. Older woody growth contains lignin, an indigestible form of fibre. As with people, the quality of diet is important to the health of the sheep.

A final note concerning sheep digestion—no matter how still the day, it is always windy around sheep. The cellulose material mentioned above is digested by the process of fermentation, a procedure which generates copious quantities of methane gas.

It isn't just the presence of four stomachs that makes sheep interesting to study; their long relationship with *Homo sapiens* is an engrossing trial of symbiosis. The domestic sheep, *Ovis aries*, can be linked archaeologically to human sites in the south-west of Asia from about 5000 BC onwards. Not long after this time sheep appeared in early wall paintings in Egypt. Humans must have introduced sheep to Egypt, because they were not indigenous to the region. Although our interest lies in sheep as wool producers, their domestication led to sheep being used for meat, milk supplies, as beasts of burden and as a source of hides. In many countries people eat cheese made from sheep's milk. Lanolin produced by sheep is a foundation stone of the cosmetics industry. Your tennis racket may even be strung with gut from friendly neighbourhood sheep.

Depending on the author one reads, there are anywhere from forty to more than 900 breeds of sheep. Wild sheep have been described as 'high-spirited, daring and self-reliant'. In contrast to wild sheep, domestic varieties have been labelled 'wary, typically timid and defenceless'. Just some of the breeds around the world include the Navajo of the United States, Zeta Yellow of Yugoslavia, Dorset Down of Britain, Kazakh Fat-rumped of Russia, Blue-faced Maine of France, Beni Guil of North Africa, Kuka of Pakistan and Hu-yang of China. The colourfully named African varieties include Congo Long-legged, West African Dwarf and Black-headed Persian. Locally, the Corriedale of New Zealand and, the most famous of them all, the Australian Merino, round out our sampling.

Structurally, sheep have certain traits resembling those of their human mammalian cousins while in other ways they are almost unique. Adult sheep vary in size from 35 to 180 kilograms, with less weighty breeds being by far the most common. They possess—from above down or front to back, pick your formula—cervical, thoracic, lumbar, sacral and coccygeal vertebrae, as do humans. Instead of twelve pairs of ribs they have thirteen pairs; perhaps the extra ribs could help support their four stomachs. They walk on 'feet-hooves' that have only two toes.

The mature sheep has thirty-two teeth, arranged in a peculiar configuration. Eight teeth are incisors or cutting teeth. All of these are in the lower jaw at the front of the mouth. On top is a ridged cartilaginous pad made from tough fibro-elastic tissue. The sheep grips grasses or other food between the lower teeth and upper pad, partially biting and partially tearing away the herbage. Such a system enables sheep to take off grasses much closer to the ground than can cattle.

Wool

We are used to thinking of sheep as producers of wool. Many breeds, especially primitive breeds, do not grow wool at all but generate a coat of hair. The skin, the place where it's all happening in the matter of wool growth, is two millimetres thick. Just like its human counterpart, sheepskin contains sweat glands, sebaceous or oily glands and the follicles which grow hair (oops!) wool. The surface area of the skin of a Merino can be rounded off to about one square metre; in terms of mass, this amounts to approximately 5 per cent of the animal's live weight

Within this area of high activity we find that the skin of a single sheep can contain from 30 million to 100 million follicles. A figure easier for the mind to grasp is 9000 wool fibres per square centimetre.

This gives one an idea of the fineness of wool. An individual wool fibre measures from 10 to 70 microns in diameter; for individuals still attuned to imperial measurement, a micron is 40 millionths of an inch.

Wool grows twenty-four hours a day and the immense number of fibres leads to a single sheep

AN EXAMPLE OF THE HEAVILY WRINKLED VERMONT SHEEP WHICH WERE IMPORTED INTO AUSTRALIA IN THE LATE NINETEENTH CENTURY. THE HARM THEY CAUSED THE MERINO BREED WAS ALMOST IRREPARABLE. *NATIONAL LIBRARY OF AUSTRALIA.*

COUNTING-OUT PENS ARE FILLED WITH FRESHLY SHORN SHEEP NEAR THE END OF A RUN.

growing over 8000 kilometres of wool in a good year or, stated another way, one kilometre of wool per hour, awake or sleeping.

Keratin, a tough sulphur-containing protein found in human skin, hair and nails, is also found in wool. Nina Hyde, *Washington Post* fashion editor, writes that keratin is 'the same protein that makes wool irresistible to moths'.

Sheep commonly live for eight to twelve years but some have survived for as long as twenty years. Males often mature earlier than females, sexual maturity in Merino rams taking place at six to seven months of age. Pregnancy lasts close to five months, the usual variance being 140 to 160 days. The breeding season takes place from late summer to winter as far as the ewe is concerned. It is thought that the breeding season is regulated by the length of the day.

The fleece protects a sheep from extremes of climate, keeping it cool in summer and warm in winter. Ninety per cent air and 10 per cent wool, the fleece is an excellent insulator to the animal harboured within. When the temperature reaches 38°C, not at all uncommon over much of Australia in summer, the temperature on the sunny surface of an unshaded Merino fleece is 85°C. The protected underbelly fleece reads 63°C but the sheep remains fairly comfortable. That's because the temperature 2.5 centimetres deep in the fleece is 49°C while at the skin level the temperature is 42°C. These wondrous insulating qualities are effective at both high and low temperatures, allowing the sheep to survive over a wide range of climatic conditions. Still, extreme heat stresses the animal and—like a dog—increased respiration in the form of panting is necessary to aid in cooling. Freshly shorn sheep can suffer from sunburn, particularly in the vast open areas of the continent where shade is quite often nonexistent.

Sheep Behaviour

Sheep feel safe in a group and it is natural for them to flock together. They are also 'following' animals. The flocking-following urge is used to advantage in sheep-handling procedures. On country properties one often sees the station hands driving a flock of sheep

A TIME EXPOSURE REVEALS SHEARERS' ARMS MOVING WHILE THEIR BACKS REMAIN STRAINED IN A BENT-DOUBLE POSITION.

DAY OLD MERINO LAMBS AND THEIR MOTHERS.

AFTER EACH OF THE DAY'S FOUR RUNS IVAN COUNTS THE SHEEP SHORN BY EACH OF THE MEN. THEIR EARNINGS ARE DIRECTLY DETERMINED BY THE NUMBER OF SHEEP THEY SHEAR.

before them. This is nowhere near as effective as having the sheep follow along. Confirmation of this fact came from Ron Taylor, longtime New South Wales Hunter region pastoralist, who spoke of men moving sheep in earlier days:

> Six men could move 10 000 sheep then. One would be in the lead, one behind and the others on the wing. The fellows on the wing would have dogs. You got the sheep in a long string; you didn't travel them all together. If you had 10 000 sheep and you got them stringing for two miles (3.2 kilometres) you'd get along better than having them all circling in around one another. You string them out. And where they've got roads to follow they generally follow along well. You trained them to follow if you didn't have any dogs with you. In fact, you trained them not to be pushed but to follow. The longer you can get your line the easier it is to move stock. If you had any sizeable number of sheep you could muster them in the morning and move them four or five miles in the afternoon.

A variation of following behaviour is commonly exploited by European shepherds. A trained sheep, referred to as a 'Judas', is utilised to lead the mob ahead. This contrasts with the usual Australian practice where the mob is forced forward from behind.

Sheep display three kinds of aggressive action: head butting, kicking with a foreleg, and pushing with a shoulder. Even so, they are non-combative to the point where they are unable to defend themselves against predators such as dingoes or wolves. Because they are so defenceless they must be constantly watchful. Fortunately, sheep have excellent vision. In a paddock they are able to see a dog over 500 metres away. Their field of vision is considerable, thought to be nearly 270 degrees with only a backwards blind arc of 90 degrees. This problem is solved by an ability to see rearwards through their back legs when the head is down, grazing. Sheep have exceptional stereoscopic vision, a capability which would help to explain their renowned surefootedness.

Sheep possess acute hearing. Station hands always seem to be yelling at them during yarding and penning procedures and I've wondered whether the sheep were impressed by all the hubbub. It seems they are not, if authors A. Barber and R. Freeman are as accurate as I suspect:

> Experienced stockmen use low volume sounds to encourage sheep to move rather than loud alarming clashes and bangs. Grandin (1983) states that all types of livestock react negatively to the sound of people yelling. A skilled, quiet stockman who makes only a small 'Sshh' noise can move more animals per hour than one who shouts.

The intelligence of sheep is frequently called into question. Certainly they are not the most intellectually gifted creatures in the animal kingdom. On the other hand, tame sheep can be trained to do all sorts of things, much like a dog. They learn by repetition but they have better memories than dogs. Before you get all enthused about a pet sheep you will want to know that they cannot be house-trained.

Merinos

Present-day domestic sheep have descended from two main groups of wild sheep. One is the Mouflon and the other is the Urial, respectively *Ovis musimon* and *O. orientalis* for the pedant. The Mouflon appears to have originated in southern Europe; the Urial in Asia.

There is no general agreement on how sheep progressed from this stage to that of fine-wool producers in Spain over a thousand years ago. A logical hypothesis acknowledges that North African sheep husbandry was found to be allied to the methods later observed in Spain. Quite possibly, invading Moors of the eighth century brought their sheep and breeding practices with them when they took Spain. In any event, under the Moors both animals and country advanced greatly. Spain became a leading power and sheep increased to several millions in number. Tillage and industry similarly progressed apace. Then, near the middle of the thirteenth century, Ferdinand wrested Spain from the Moors and it was all over bar the shouting for ruminants and residents. At one stage the number of Merinos declined from 7 million to 2.5 million under the unenlightened management of the Christians of the day.

No matter the government, sheep were always important to the nation's economy. There were two main classifications of flocks, the *Merino transhumantes* and the *Merino estantes*, indicating simply travelling flocks and estate–or fixed grazing—flocks, respectively. Certain owners, that is nobility and favoured individuals, were allowed to graze their animals over large areas of the countryside as the travelling Merinos journeyed to the mountains each spring and back down to the plains before winter set in. Great wide drovers' roads ran across Spain and, naturally enough, the sheep had to feed as they moved along. The king made laws forbidding farmers to fence their fields or even to drive the sheep off if they fed on the farmers' crops. Land was commandeered as needed and protesting farmers could be put to death.

Extensive tracts of forests were cut down from the mountains, converting sizeable areas of Europe's alpine and lower levels into grazing lands for hungry sheep, and devastating the natural environment. Because the Spanish king possessed vast holdings in Italy, he repeated the process in that country. At the time this avarice created considerable wealth for the king and his cronies, while it brought privation to the unfortunate peasants. In the end the loss belonged to all of Spain.

The origin of the word 'merino' is as uncertain as that of the beasts themselves. It possibly comes from a Spanish word for 'thick, curly hair'. Or from a Spanish word meaning 'moving from pasture to pasture'. Maybe it refers to the Sierra Morena, high country where the sheep pastured in summer. Or Beni-Merines, the name for one group of North African Berbers who participated in the invasion of Spain. Could it be from a well-known Moroccan family named Merino? Take your pick.

At this point one may ask, 'Why all the fuss over the Merino when there are hundreds of varieties of sheep available?' The answer to this question puts everything else into perspective. Money, power and prestige, the engine drivers of human endeavours, are intimately involved in the history of the Merino.

The breed is a superb producer of wool, both in quality and quantity. Its wool is finer than that of many other breeds. But the salient feature of the Merino is its skin follicles. Wool grows up and out of these skin follicles like plants growing up and out of pots. There are secondary follicles surrounding primary follicles and each of these secondary follicles supports a fibre of

The Albury Ram

As I was going to Albury along the other day,
I saw the finest sheep, sir, was ever fed on hay

Chorus:
Singing blow your winds to morning, blow your winds, hi-ho,
Blow away the morning dew, blow boys, blow.

The sheep he had four feet, sir, upon which he used to stand,
And every one of them, sir, it covered an acre of land.

The sheep he had two horns, sir, they grew so mighty wide,
They're going to build a bridge with them from Derbyshire to Clyde.

The sheep he had a tail, sir, it grew so mighty long,
'Twas used to build a telegraph from Sydney to Geelong.

The wool upon his belly, it bore him off the ground,
'Twas sold in Melbourne the other day for a hundred thousand pound.

The wool upon his back, sir, it grew so mighty high,
The eagles built their nests there, for I heard the young ones cry.

A hundred gallons of oil, sir, were boiled out of his bones,
Took all the girls in Albury to drag away his frame.

The man who owned this sheep, sir, he must have been mighty rich,
And the man who made this song up was a lying son-of-a-gun.

wool. Most other breeds have from three to seven secondary follicles for each primary. The Merino has twenty to twenty-five secondary follicles, turning it into a veritable wool factory. Drawing the simile with plant life once again, it's like watching a forest of bamboo grow alongside one of oak.

Nina Hyde notes that wool was 'the commodity that brought wealth and power to England for 700 years. Wool was England's first great industry. By the late Middle Ages its export had become the nation's largest source of income ... every European country relied on England for it.' Across from the British Isles on the European continent, Spain had a law forbidding the export of even a single Merino. The government meant business: the penalty for taking a Merino out of its homeland was death. Even with this Draconian measure the English had acquired 3000 Spanish Merinos late in the fifteenth century and they repeated the feat about a hundred years later, all to no avail as far as their own breeding stock was concerned.

The Spanish maintained separate flocks of Merinos which possessed different features. The *Paulars*, named for an order of monks, were handsome and fine woolled. Some of their characteristics are noted in today's Australian Merino. The *Negretti* were tall and strong with a tendency to be wrinkly. The *Escurials* were very fine woolled and used in development of the Saxon variety. The *Infantados*, owned by the Duke of Infantado, were shipped in part to America, France and Saxony. All of the foregoing are believed to have influenced Merino development in Australia.

NATURE-CAN-BE-HARSH DEPARTMENT. THIS PAIR SUCCUMBED TO FLY STRIKE. SHEARERS REFUSE TO SHEAR FLY-BLOWN SHEEP UNLESS THE SHEEP ARE FIRST TREATED BY THE STATION HANDS.

England grew its own coarse-woolled sheep while depending on Spain for fine wools. The Elector of Saxony (part of Germany) was a cousin to the Spanish king who, in 1765, relaxed the ban on Merino exports and sent 300 to his relative. Saxony's economy had been shattered by the Seven Years' War and this was a measure to aid its recovery. In 1786, Louis XVI purchased almost 400 ewes from Spain, housing them at Rambouillet Park which he created as France's national stud. The sheep sent to England, France, Germany, and later America, were tied to the development of the Merino in Australia.

When Napoleon's armies invaded Spain, 20 000 Merinos were sent to France as spoils of war. Next, the Duke of Wellington came to Spain and thousands of Merinos ended up in Britain. Spain's reign on the plain had ended.

Returning to the British situation, we find that commerce in wool was a leading English industry right up to the nineteenth century. Because wool was vital in providing jobs at home and as an instrument of government policy overseas, it was an industry under tight rein, directed originally by guilds and later by the state itself.

For a very long time, bringing woollen goods into Britain was prohibited. The government further aided the wool manufacturing industry by making the export of sheep a highly illegal activity. Incidentally, the penalty for an exporting first offender was one year's imprisonment plus the cutting off of his left hand on market day, the hand to be nailed up in a prominent position. The second offence earned the death penalty. Neither compassion nor leniency featured in the strategies of the authorities in the good old days.

It can be seen that the mother country was looking out for her own interests. The Great Southland was a colony and it was intended to remain that way; even in discussion of woollen cloth, 'the finest quality' was to remain in Britain while New South Wales manufacturing 'should be confined to the coarser kind of cloth'. The romantically named East India Company was a trading giant, controlling all Indian Ocean and Pacific Ocean commerce via royal charter. Fortunately for the new land, ways and means were found to loosen the reins which extended across 16 000 kilometres of open sea.

One of these 'ways and means' took place in 1793 when the *Shah Hormuzear* discharged one hundred Bengal sheep at Port Jackson. These sheep, as well as a few others, reproduced at an amazing rate, so that by 1800 there were 6000 in the colony. They weren't Merinos as we know them today. The first Merinos landed in 1797 and it is generally conceded that even these had 'cross' blood in them. They numbered thirty and had been purchased in South Africa from the estate of the late Colonel Robert Gordon, a Dutchman of Scottish lineage. It was rank good luck that found three ships seeking colonial supplies just as

A MOB OF 5000 YOUNG MERINOS BEING MUSTERED FOR CRUTCHING.

Colonel Gordon's Merinos were being disposed of by his widow. According to current reckoning, no more than thirty Merinos entered Australia prior to 1820. In the following decade many more came and it is from these later imports that the present Merinos have evolved.

One of the major Merino importers was the Australian Agricultural Company; in this cosy arrangement, about 400 000 hectares were given to a pastoral company that happened to have as some of its shareholders, thirty members of the British parliament. The shipping lists detailing Merinos brought to Port Jackson for the Australian Agricultural Company are most evocative. Here are some highly abridged examples:

> The *York* sailed in June 1825 with 313 French Merino ewes, 12 Anglo Merino ewes and 15 French Merino rams. Also on board were 1 Durham bull, 1 Durham cow, 5 Scotch Highland cows and 1 Scotch bull.
>
> The *Prince Regent* (under Captain Lamb!) sailed in August 1825 with 215 Saxon Merino ewes and 25 Saxon Merino rams.
>
> The *Elizabeth* sailed in November 1826 with 60 Merino and Leicester sheep. Also, 1 cow.
>
> The *Marquis of Anglesey* sailed in June 1827 with 275 French Merino ewes and 9 French Merino rams. Also, 10 Welsh ponies.

A tragic tale accompanied the delivery of some Saxon Merinos, shipped from Germany to Sydney in 1824 with Captain Cairnes of the *Cumberland*. All went well and the master made a return trip, this time accompanied by his son and daughter. Almost home, pirates captured the ship. Captain Cairnes was ordered to walk the plank and as he started off he clutched the girl, taking her with him into the briny. The boy jumped in after his father and sister, concluding one ill-fated drama in the theatre of wool.

The passage of time brings us to a famous actor on wool's stage, a man who has been described as everything from a farce to a tour de force. John

Macarthur landed in Sydney in 1790, a brash young military officer soon to become an ascendant figure in the colonial Rum Corps. A splenetic individual, he was a perennial combatant; nevertheless, he enjoyed the largesse of two early governors of New South Wales as well as patronage from far-off Britain. With both land and convict labour freely supplied to him Macarthur went on to become the colony's chief sheep-owner. His belligerent temperament frequently landed Macarthur in strife, to the extent that he spent many years back in Britain, in exile of one form or another. Even this apparent misfortune was really a blessing in disguise. It allowed Macarthur's wife, Elizabeth, to get on with the task of running their sheep properties. Elizabeth was an extremely astute woman and the Macarthur Merinos flourished under her charge. It is freely acknowledged today that much of what the Macarthurs accomplished in establishing Merinos in Australia was achieved by Elizabeth, not John.

Of late, he has been described as a notorious monopolist, a man without scruples, a person possessing the traits of a confidence man. Author Robert Hughes says simply, 'He died mad.' Macarthur's picture appeared on the now discarded two-dollar bill. That was before historians gave him a thumbs-down verdict. Macarthur seems to be one of those individuals who attain fame and fortune in spite of himself. Whether history ultimately credits John, Elizabeth or fortuitous circumstances, the Macarthur name invariably will be associated with the development of Merinos in Australia.

Did you know that in the late 1800s the Americans invaded Australia? It was just one group of Americans that were aggressors and it was just one area of endeavour that was affected: sheep breeding. Surprisingly, this welcome invasion from Vermont destroyed certain sections of the local stud business and came close to irreparably harming the Australian sheep wool industry at the turn of the last century.

In the 1860s, a Vermont sheep breeder created what were then reputed to be super-sheep because of the tremendous quantity of wool they carried. Just a few were brought to Australia and at Wanganella, one of the nation's leading studs, they assisted in creating a greatly improved local Merino. Too bad well-enough was not left alone. Late in the century the invasion began in earnest. By then the Vermont rams had become exceedingly wrinkly and these folds allowed an amazing growth of wool to take place. The imports leapt in popularity among many breeders as wrinkles became fashionable. A portion of the pastoralists, however, didn't take to these super-sheep and a bitter debate ensued.

The positive side to the Vermont sheep was the huge amounts of wool carried on each beast. The negative side was much stronger: the wool was exceedingly greasy and when scoured the net weight became markedly reduced. The wool was tainted by a sprinkling of hair, bringing forth unbridled sarcasm from Thomas Shaw, a man widely acclaimed for improving Merinos—and thus wool—in Australia. Shaw said that these new sheep seemed to be growing cow hair. 'So the next step, I presume, will be to grow wool on cows.' And these criticisms were only the beginning. Vermont sheep did poorly in local drought conditions. Their grossly wrinkled bodies became ready targets for blowflies, and shearers were sorely troubled by the task of shearing amidst the wrinkles and heavy grease. For a few years, the fashionable Vermont-bred sheep led the winners at all the major sheep shows. Nevertheless, when their disadvantages came to the fore it marked a quick ending to a brief, bitter struggle. By 1905 the war was over, with commonsense in breeding being the ultimate winner.

The Peppin family, early owners of the famous Wanganella stud property, used Rambouillet, Negretti, early Vermont and, it is hinted, even a touch of Lincoln bloodline in creating the Peppin Merino during the late 1800s. Peppin bloodlines are well regarded to this day, their progeny highly numerous throughout the country.

A South Australian Merino variant was developed in response to the vast and arid nature of grazing land in that state. The South Australian Merino is big and long-legged for it often must walk long distances to reach water. Two brief anecdotes concerning South Australian stud pioneer, Mr John Hughes, show the tenor of the man. During early breeding experiments he noted that 'many of my sheep bore a strong resemblance to young camels'. On another occasion he was visiting William Macarthur—son of the celebrated John Macarthur—at Camden Park. The host gave Hughes piles of information concerning his flock and the visitor enquired as to why they used such sheep when his own returned approximately twice the amount for each fleece. Macarthur said, not without pride, that they bred for pure blood. Hughes responded that the South Australian way was to breed for pure money.

Another local variant is the Saxon Merino, developed from sheep brought from Saxony, some blood extending back to the Spanish *Escurial* flock. Saxons are smaller than the robust Peppins. However, their wool is extremely fine and highly regarded. These sheep do well where rainfall is greater, for example in Tasmania and the Mudgee district of New South Wales. Finally, the Spanish Merino is so named because of its lineage. They are the least numerous recognised group, existing in Australia's more parched regions.

The modern Merino is described by E. W. Cox (in R. Anderson's *On the Sheep's Back*) as:

> A large-framed sheep, symmetrical in form, with a wide developed neck, deep body and well-sprung ribs—short legs, a straight back, and in appearance is almost rectangular. It is a robust, hardy sheep bred to stand squarely on all four legs. The wool is closely packed on the whole frame and is uniform and even.

It wasn't so at the outset. The nation began with sheep which were, according to Alfred Hawkesworth, 'thin, narrow, long-legged, flat-sided, razor-backed sheep, with covering more like hair than wool'. Those sheep which were pastured around colonial Sydney never had a real chance; it was many years before residents realised that the moist coastal environment was deleterious to the health of sheep. Due to the barrier of the Blue Mountains, the local settlers allowed themselves to be cramped along the eastern seaboard.

Governor Macquarie commented: 'It must be a matter of astonishment and regret that amongst so large a population no-one appeared within the first twenty-five years, possessed of sufficient energy of mind to induce him to fully explore a passage over these mountains.' Then, in 1813, as all Australian school children know, Blaxland, Lawson and Wentworth found a way through the Blue Mountains and an immense area of magnificent grazing land was opened up. Settlers poured in and Australia began a journey that continues to this day. She is still a nation that rides on the sheep's back.

Prior to the Napoleonic Wars, most English wools of high quality came from Spain. After these conflicts the German-Saxon wool led Britain's imports. Problems arose as they tried to breed even softer wool; in the end the German sheep were weaker overall. There's no doubt that this enervation of the strain provided an ideal opportunity for Australian woolgrowers to corner a large portion of the British market.

It was a situation where a colony was established for the modest goals of providing naval supplies for the East India Company and serving as a 'dump' for criminal elements. Lo and behold, that colony came to be a source of riches for the mother country.

Sheep numbers reflect the growth of the local wool industry. In 1800 there were 6100 sheep in Australia; in 1821 there were 290 000; by 1843 the number had leapt to nearly 5 million. By 1891 more than 60 million sheep grazed over Australia and today there are ten sheep for every one of 18 million Australians. What began as a depository for living trash has become the globe's greatest producer of wool, today accounting for fully 70 per cent of the clothing wool used worldwide.

Captain Cook's sailing colleague, Sir Joseph Banks, renowned for his thoughtful observations, was wide of the mark in assessing Australia's sheep-raising potential:

> I have no reason to believe, from any facts that have come to my knowledge, either when I was in that country or since, that the climate or soil of New South Wales is at all better calculated for the production of fine wools than that of other temperate climates, and am confident that the natural grass of the country is tall, coarse, reedy and very different from the short and sweet mountain grass of Europe, upon which sheep thrive to best advantage.

Little did Banks realise that this wide brown land which suffered from the tyranny of distance, extremes of climate and harsh masters overseeing a recalcitrant workforce, would literally bloom into an ideal home for the Merino. Along the way the cream of the early pastoralists became known as 'pure Merino', a label which identified the traditional, conservative aristocratic colonials. It was an ironic turn of phrase based upon the timid, witless, unpretentious cross-bred Merino sheep which was to elevate its host nation into resounding prosperity within a brief span of years.

IVAN LETCHFORD, SHEARING CONTRACTOR, WITH SMILEY. THE TRUCK IS LOADED AND READY TO HEAD FOR THE BUSH.

A Day in the Life

The line of the horizon is banded with the sublime red of the Australian dawn. By the time the sun has poked its head up, Ivan Letchford's head has also lifted from its pillow on an old and saggy bed in a shearers' hut.

His grey hair is thinning now but the eyes remain as blue as ever. The body is fit, spare but not lanky, weighing 83 kilograms and measuring 185 centimetres in height. The voice is never soft or harsh but it can be quite firm. It is not always easy to discern the twinkle in his eye when he looks directly at you and, poker-faced, relates the most far-fetched tale this side of Johnson's Snakebite Antidote.

Ivan Letchford is a shearing contractor—has been for nearly forty years—and, depending on the season, he bears a working knowledge of the day's dawning. Shearing used to be done for just a part of

DAWN ALMOST ILLUMINATES THE ENTRANCE TO BANGATE, A NORTHERN NEW SOUTH WALES SHEEP PROPERTY. IN 1990 HEAVY RAINS TURNED THE PROPERTY INTO A QUAGMIRE WHERE THOUSANDS OF YOUNG SHEEP DIED.

HIGH COUNTRY PROPERTY DURING THE SHEARING SEASON.

the year. Spring was the time favoured by the landowner and the shearing season continued for several months as the weather gradually warmed. Now the 'season' lasts all year, with the crew returning to work after the Christmas holidays and then sticking with it until the following December.

There are two kinds of shearers, according to their method of work. The suburban shearer is a latter-day development, a man who lives at home and travels to surrounding properties to do the shearing. The traditional shearer, Ivan Letchford style, is in declining numbers. The traditionalists are the elder statesmen of the profession, ranging over the countryside from property to property, a cook accompanist looking after the victuals, the group moving to the next shed only when the present job has been completed.

After morning ablutions, Ivan and his crew sit down to a hot breakfast at 6.30 a.m. It is the first of five meals consumed each working day; three being hot and two cold. These last two are the morning and afternoon smoko breaks. Shearing is extraordinarily hard yakka and the energy requirements to sustain a man working with robust struggling sheep all day are such that five meals are just about the right number.

For breakfast Ivan likes chops, 'bubble and squeak' (fried up vegetables left over from the previous day) and toast. 'The toast must be cooked over a wood stove to develop decent flavour', is Ivan's verdict on its preparation. The chops are mutton, as is almost all the meat eaten by the crew. The property owner supplies the meat and you can safely bet that veal scaloppine or barramundi steaks won't be on the menu. In their place look for baked mutton, curried mutton, marinated mutton, mutton steaks, mutton meatloaf, corned mutton, mutton sausages or even mutton-based spaghetti bolognaise. One or more sheep are butchered each evening, the size of the crew determining the number required for tomorrow's meals. The carcases are hung in the meat room overnight to be broken down to cooking size the following morning.

Shearing is done in two-hour segments, the early morning commencement scheduled for 7.30 a.m. Before the start, Ivan checks to see that the sheep have been correctly yarded for the day's shearing.

Click Go the Shears

Out on the board the old shearer stands,
Grasping his shears in his thin bony hands;
Fixed in his gaze on a bare-bellied yoe
Glory if he gets her, won't he make the ringer go.

Chorus:
Click go the shears boys, click, click, click,
Wide is his blow and his hands move quick,
The ringer looks around and is beaten by a blow,
And curses the old snagger with the bare-bellied yoe.

In the middle of the floor in his cane-bottomed chair
Sits the boss of the board with his eyes everywhere,
Notes well each fleece as it comes to the screen,
Paying strict attention that it's taken off clean.

The colonial experience man, he is there of course,
With his shiny leggin's on, just got off his horse
Gazes all around him like a real connoisseur,
Scented soap, and brilliantine and smelling like a whore.

The tar-boy is there waiting in demand
With his blackened tar-pot, in his tarry hand,
Spices one old sheep with a cut upon its back
Hears what he's waiting for it's 'Tar here, Jack!'

Now the shearing is all over, we've all got our cheques
So roll up your swags and it's off down the track,
The first pub we come to it's there we'll have a spree
And everyone that comes along it's 'Have a drink with me'.

There we leave him standing shouting for all hands,
Whilst all around him every 'shouter' stands,
His eye is on the keg which now is lowering fast,
He works hard, he drinks hard, and goes to Hell at last!

This has to be done at least four hours in advance, or if there are ewes with lambs afoot the time for yarding is reduced to two hours. If sheep are hot they tend to be overactive, increasing the chance that they will struggle and kick the shearer. If sheep have a bellyful of water or food, it makes shearing uncomfortable for man and beast.

Most stations have engines of 6-20 horsepower (petrol or diesel) for power around the shed and shearing team's quarters, the engine size depending on the number of shearing stands each engine must drive. This machinery is looked after by a person called the 'expert'. Besides attending to the power plant, the expert grinds the combs and cutters for the shearers, adjusting the handpieces if necessary. When he has a relatively small team of shearers, say six or eight men, Ivan is his own expert. So, by the 7.30 a.m. start he has checked the engines and fuel, gone over the machinery and had a chat with the presser, after a fleeting inspection of the bales of wool.

From 7.30 to 9.30 a.m., the shearers work at full tilt. Briefly, the board boys pick up the fleeces and deliver them to the classing table; the wool rollers remove burrs and vegetable matter from the raw wool and the classer then grades the wool. It is next placed in one of several bins from where the presser makes it into 180 kilogram (400 pound) bales. Again, depending on the size of the crew and the job at hand, Ivan may act as his own wool classer.

It's 9.30 a.m.—smoko! No sheep are caught after the ringing of the three-minute bell at 9.27. This gives the shearers time to complete the last sheep and allows the board boys and rouseabouts time to clean up, thereby enjoying a full thirty-minute break. The cook has already brought morning tea to the shed. It commonly consists of a variety of sandwiches with cakes or biscuits, and a large pot of tea carefully prepared with rainwater. Before his own break, Ivan retrieves the all-important tallybook and takes it to the counting-out pens where he counts the sheep each man has shorn during the previous two hours. When he returns to the shed the tallybook is usually gone over with considerable interest by the shearers. Naturally enough, they want to see how they—and their colleagues—have done during the day's first run.

To me, the most obvious change brought about by smoko is the instant quiet which pervades the shed. As the machinery, shears and press are stilled, the welcome calm of the Australian bush seems to settle onto man and beast. I never cease to be entranced by all the activity during working periods in the shed, but it is the break periods which I particularly relish. The instant quiet, the perceptible relaxation, the breaking of food, the wide-ranging conversation, the good-natured banter, all these combine to say, 'Repeat at regular intervals as an aid to mental health'.

By 10 a.m. the expert has the machinery running once again and also has ground the shearers' combs and sharpened their cutters. A fine edge is necessary to cut the wool away cleanly and swiftly. 'The cutters have four teeth which oscillate 3000 times per minute across the face of the comb,' Ivan relates, 'and if they get away they can cause nasty wounds to sheep or shearer.'

IN THE EARLY STAGES OF SHEARING A SHEEP. THIS SHEARER IS 'OPENING-UP THE FIRST HIND LEG'.

The next two hours are a repeat of the first, the objective being to shear as many sheep as possible with the least amount of damage and lowest expenditure of energy for all participants. At midday, Ivan retrieves his tallybook once more and totals up the sheep shorn before heading off for lunch. The cook always has a baked dinner. Surprise–it's mutton! The shearers pay for their own sustenance (with the exception of the meat provided by the property owner) and a decent cook is high on the priority list for all of the crew. Ivan points out: 'Cooks are judged by their ability to provide good food, with variety, at an economical price.'

A few minutes before 1 p.m. the engines are started for the two afternoon runs: 1-3 p.m. and then 3.30-5.30 p.m. An eight-hour working day is spread over ten hours. It is more shearing, wool rolling, classing, pressing and tallying. At the end of the working day it is time to shed the sheep. If there are twelve or fourteen shearers this creates too much work for the penner-up to do alone, so Ivan gives him a hand. Anywhere from 600 to 2500 sheep are put under cover so the shearers do not lose a day's work if it rains during the night. The men won't handle wet sheep because they believe that doing so causes illness.

Shearers and shedhands are very clean people, and by 6.30 p.m. or thereabouts all and sundry have managed to enjoy a shower. A beer or two may relieve parched throats, for beer is tolerated in the evenings as long as it does not affect the next day's performance. On quiet evenings Ivan will total the day's tallybook, update the woolbook and perhaps send a report along to the selling broker. Depending on the locale being worked, he may enjoy a pleasant dinner and social evening with the grazier and his family, just once or as often as several times during the week. He may even see other owners that he works for in the surrounding area. Often Ivan gets invited out if it is known he is in the district. An evening of relaxed companionship and an opportunity to enjoy creature comforts presents a welcome diversion after an arduous working day.

Who is Ivan Letchford and why has he spent his entire adult life working in the world of sheep?

Ivan is a Sydneysider, born in Liverpool prior to the Great Depression. His father had a combined General Motors-Holden agency and mechanical engineering business in Cabramatta, an undertaking which went downstream during the Depression. For a few years, the family moved according to the dictates of the times and then, aged fourteen, Ivan left school to attend a course in 'sheep and wool' at East Sydney Technical College. It was a three-year course during which the first six months of each year were spent in formal education and the second six months involved practical experience in the bush. Ivan says, 'I always found the course extremely interesting and when I got to the shearing shed I liked the work immediately. The senior men in the shed assisted in teaching me practical ways of going about my work.'

Working in the Mungindi area of New South Wales on his first job, there was the daunting number of 38 000 sheep to be shorn. 'I was always proud of the fact that I returned to that shed two years later in the capacity of a wool classer—a very responsible job.'

Shearing time is literally harvesting time for the grazier. The efforts of the previous twelve months culminate in the brief but intensely active shearing season. Ivan's first classing job took place when he was aged seventeen, a time when wool cost 11 or 12 pence per pound. A ha'penny in either direction was terribly important to the landowner's annual income. Wool he had classed sold extremely well, which was a good reference for the eager youth. At this period in wool's development it was not auctioned. Instead, the fibre was appraised by two representatives of the Australian Wool Board and the selling broker. The federal government was maintaining quantities of wool throughout Australia because the effects of World War II had not yet been overcome. Ivan held the appraisal records for wool—49–½ pence per pound—for a duration, and he became known as an up-and-coming young classer.

TWO JIMS.

JIM REES HAS BEEN SHEARING SINCE 1962. ON HIS FIRST DAY ELEVEN SHEEP WERE SHORN IN EIGHT HOURS; JIM'S BEST DAY WAS 221 FLEECES SHORN.

JIMMY 'THE RED FELLOW' BRAYSHAW, SHEARING TEAM 'EXPERT', RELAXES WITH A COOLING DRINK OF WATER DURING A SMOKO BREAK.

A KEEN OBSERVER OF HIS SURROUNDINGS, THIS YOUNG KELPIE PUP HAS ALREADY BEGUN TRAINING TO BECOME A WORKING SHEEP DOG.

Endeavouring to learn as much about the industry as possible, he worked in woollen mills, then in the waste trade where reclaimed fibres were utilised anew. At Botany he worked in a wool-scouring plant, removing grease and dirt from the fibre. Another job was concerned with the classing of sheepskins.

This same company also dealt in rabbit skins and Ivan remembers instances where small pelts were piled solidly in heaps 30 metres long and 15 metres high. 'Of course, during those years in Sydney, the 'rabbito' came around in a horse-drawn cart, calling out 'rabbit-oh, rabbit-oh', purveying what was then a popular meat because of its affordability for the average family– costing 10 or 11 pence each.'

Warming to the discussion of rabbits, Ivan's view is that the rabbit did a great deal of damage to the wool industry early in the century. Rabbits ate all the grass, they ate the bark off the trees and they killed off the edible scrub, which was the greatest standby a grazier had during a drought. An enormous expense was necessary to get rid of them.

'An illustration of the number of rabbits we are discussing comes vividly to mind. Friends on a rabbit-infested property had to fence off the water tanks at night to prevent the blighters from getting into them. One Saturday evening they went out, noticing half-a-dozen trapped rabbits. When they returned, only a matter of hours later, they had to kill 3500 rabbits! I know of real tragedies on properties around 1900: one year 80 000-90 000 lambs may have been marked while the next year, there were but 3000 lambs marked.'

Ivan had a friend in the Moree district of New South Wales who had difficulties in obtaining shearers for his property. It was suggested to Ivan that he secure a team and he did so, the result being quite satisfactory all round. 'I found the job extremely enjoyable and challenging. I really liked all the characters I met, the humour was refreshing and I found that I was able to cope well during the frustrating times.

'It was about 1954 that, in Sydney, I approached the managing director of Glen Rock station seeking a contract for their shearing. They were running about 50 000 wethers then. I got the job and have been going there every year since, except when British Tobacco owned the property and they ran cattle, not sheep.' Glen Rock is a beautiful pastoral property north-east of Scone, near the headwaters of the Hunter River, within the rugged terrain of the Great Dividing Range.

After the initial contract work done at Glen Rock, Ivan approached various graziers with a view towards establishing his own business. He secured what appears to be an extremely loyal clientele. The majority of property owners hiring Ivan's services have been with him for more than twenty years. He has had as few as one small team to as many as five teams, totalling thirty men, shearing at the same time. Almost all of his undertakings are conducted within the boundaries of a large New South Wales triangle, delineated by the Upper Hunter, the Walgett-north-west area and the central west. While the last-named area still holds some jobs and many friends, the region has moved towards farming, especially wheat, eroding contacts in this part of the state.

Usually Ivan drives around 2000 kilometres each week. A venerable Bedford truck on its sixth or seventh motor hauls Ivan, dogs, food, freezer, portable wool press and everything but the proverbial kitchen sink over some of the roughest, most potholed tracks the country has to offer. Weekends are a different matter.

The Letchford family home is perched on a hillside at Church Point, with what can only be described as magnificent views over the Pittwater basin. Ivan attempts to get home most weekends, missing out only occasionally when the team becomes stuck fast at a flooded or distant north-west property.

BOTH MEN ARE 'OPENING-UP THE NECK' ON THEIR SHEEP.

Ivan says, 'I've got one wife, two sons and four or five dogs, good workers the lot.' Madge Letchford is one of the best cooks I've ever encountered. A snack for two or a meal for twenty is accomplished with the same calm, knowledgeable, seemingly effortless manner.

At any given time, two or three of the canines are respectable working sheepdogs, the remainder in various stages of training or retirement.

It doesn't look like Ivan Letchford will be retiring in the near future. A recent article in the daily press features one of my friends who had sold his stockbroking firm a few months previously. The writer said that the individual continued to show up at his office each day and it was likely that he was going to have to be dragged out by his feet. I think a similar prediction could be made about Ivan Letchford: the only way he will completely forgo the pleasure of paddock and shearing shed is if he is dragged away, kicking and screaming, to the local retirement village.

2

EARLY *Days*

In the early 1800s Australia's sheep were quartered in New South Wales and Van Diemen's Land. The first local pastoralists grazed their sheep in the Sydney environs. Unfortunately, they did not realise that this region was too wet for the animals' liking. Parasites did find the area favourable and these two factors impeded development of the colonial wool industry. Another cause—this time official in nature—held back the expansion of sheep in New South Wales. The government planned to maintain control over the colony and this was facilitated by keeping everyone domiciled centrally. Once the Blue Mountains were crossed, this containment was no longer possible. Broaching the mountains opened the fertile, sun-drenched inland to settlers and sheep. Both did well in the new territory and this created a strong demand for land.

Late in the 1820s, Governor Darling tried to bring a degree of order to the homesteading boom. He had been swamped with reports concerning property disputes, bushranging, duffing, and killing of Aborigines. It was his plan to keep development restricted to what he termed 'the limits of location', a ribbon of land 240 kilometres wide, extending from Port Macquarie in the north to Batemans Bay in the south. It was too late. His successor, Sir Richard Bourke, according to Hector Holthouse, then

> authorised Commissioner of Crown Lands, whose job it was to protect Crown Lands from unauthorised occupation, to keep peace between settlers and Aborigines, to suppress sly grog-making, to keep track of ticket-of-leave convicts and to report every case of a black woman living with a white man.

Colonists simply moved onto open land and claimed it for themselves, in the process gaining the name 'squatters'. An English word, 'squatter' was used first in America where settlers squatted, or took possession of land that was in the public domain. In Australia, an official used the term in about 1828 in a deprecatory manner as he described the habit of squatting by freed convicts. He compared squatters unfavourably with 'respectable Settlers'. It wasn't long before the squatters became so successful that the term indicated notability, prosperity, dignity or, in short, respectability.

The topic of squatters and what has been termed the 'squattocracy' of Australia is an emotive one in our society. Depending on who is speaking, one is informed that squatters are either villains or heroes. One viewpoint holds that squatters simply laid claim to empty lands, used them heedlessly for whatever personal gain could be secured and often, when the earth was squeezed dry, moved on to rape a new area. Strong words indeed; words to match the vehement feelings of those opposed to the squattocracy. The case for the defence? It too is vigorous and points out that the British 'new chum' settlers knew nothing of

A PARTY OF COLONIAL SHEARERS TRAVELLING TO THE NEXT SHED ON 'THE WALLABY TRACK'. *NATIONAL LIBRARY OF AUSTRALIA.*

the country, its inhabitants or the sheep they were to introduce. They followed closely in the wake of explorers and frequently became explorers themselves. At the same time, they had to defend their stock and their families from the depredations of marauding natives who possessed intimate knowledge of local climate and conditions.

Courageous explorers or rapists of the nation, squatters did make many mistakes. As a witness at a Royal Commission stated, 'There is no doubt in my mind that the carrying capacity of the country was greatly overestimated by the early settlers'. The squatters were highly motivated by self-interest and a common fault of the group was to overstock their paddocks.

It is widely accepted that the settlers attempted to graze too many animals over delicate pasturage. Nevertheless, that common nemesis, the government, raised its ugly head just as often in colonial days as it does now. In the 1830s, upon payment of £10 per year, a pastoralist obtained a 'squatter's licence' which allowed him to graze stock on crown land. There was no security of tenure in this ruling, in fact the authorities could remove the grazier at any time without recompense. Naturally enough, this inspired the squatters to get what they could as quickly as they could without returning anything to the land.

As the nineteenth century progressed it became more difficult to enter the squatting ranks because greater resources were required. By the 1870s it was only the wealthy who could launch a squatting venture. John Ferry gave a picturesque description of the times in *Walgett Before the Motor Car* when he declared that it was 'said in those days that the first man on the land breaks his heart, the second goes into bankruptcy court and the third makes a fortune'. The third man arrived in the 1870s.

If there was one certainty in pastoral colonial life it was uncertainty. The phrase 'boom or bust' could have been invented to describe Australian rural history. Drought was as sure as death and taxes. (Dry periods have always plagued the nation: the early years of the 1980s are an excellent example.) From 1829 to 1860, eleven years of drought distressed the man on the land. Economic depressions cost many a pastoralist his property. Squatters overextended in their borrowing, banks did likewise in their lending, while money men in London tightened purse strings to choking point. Foreclosures late in the nineteenth century put properties into bank coffers and then the banks couldn't find buyers because dreadful economic and climatic conditions prevailed. It isn't often recalled that there were twenty-six banks nationwide in 1887 but by year's end in 1893 only six remained in business.

TODAY 'A SHEARER ILLUSTRATES THE USE OF YESTERDAY'S EQUIPMENT: BLADE SHEARS, WHICH REMAINED IN WIDE USE UNTIL THE EARLY 1900S.

The droughts were sometimes followed by damaging floods. Poor economic conditions stopped many public works projects, and unemploy-ment was rampant during these periods. Rabbits devastated whole sections of the countryside. Government land regulations were convoluted and open to deception. Naturally enough, swindles occurred and corruption scandals blotted the government's copybook. Prickly pear ravaged portions of the countryside and fatal disease outbreaks afflicted flocks of sheep. With such a multitude of thorny problems, overstocking was just one more obstacle to success. Little wonder that rural Australians are known for their stoicism: since day one, their lot has not been an easy one.

Before leaving this matter of overstocking, it is essential to quote C. E. W. Bean who wrote in *On the Wool Track* in 1910 about the effect of 'white man, his sheep and his rabbit' upon a delicate land that had enjoyed freedom for a million years:

Australia has been shut off from the rest of the world ever since the grasses and trees and animals began their fight for living. If anything ever needed tender, scientific handling, it was this dainty covering of grass and trees. So when the white man, raw, inexperienced, ignorant, struggled out on to those apparently rich plains and proceeded to manhandle the scrub

THE COLOUR OF THE GALAHS BREAKS THE MONOTONY OF THE INLAND.

THE MIDDLE PORTION OF BIG BURRAWONG SHED IS PART OF THE ORIGINAL STRUCTURE DATING BACK TO 1876. POSSESSING EITHER 101 OR 104 STANDS IN BYGONE DAYS, SO MANY MEN WORKED IN THE SHED ENVIRONMENT THAT A BROTHEL OPERATED SUCCESSFULLY ADJACENT TO BIG BURRAWONG.

> and grass, in less than twenty years he succeeded, too often, in destroying much of the wealth that had been gradually stored there from the beginning of the world.

So, we have before us a fresh continent. How did the early settlers go about establishing themselves on their 'runs', as properties were then called?

At first, land was simply occupied, without recourse to laws or legality. If you settled upon a tract of land, it was yours. The land was there, waiting, almost looking for tenancy. The British—for that's what the early settlers were—took possession of what they wanted, regardless of the Aborigines who had lived on (and off) the land for countless millennia. It is fair to say that they endured the usual fate of primitive people meeting a more advanced society, that is, they were seen as obstacles to be overcome. How much land one claimed often depended on how much one could control. In one area an unwritten rule maintained a distance of three miles (5 kilometres) from the homestead of the closest squatter. In a district with a stream, a similar unwritten rule allowed a new settler to lay claim to three miles of land along the waterway, in *both* directions from the homestead. Turning away from the stream he commanded as much grazing land as the eye could see.

Victoria was settled later than New South Wales or Tasmania. Part of the state was called Australia Felix, meaning 'Happy Australia', by Major Thomas Mitchell, who explored this lush portion of Victoria in 1836. New settlers simply moved beyond their predecessors and took up fresh, unoccupied countryside.

Once a holding was claimed it became necessary to erect accommodation. A typical start was to erect a bark hut which often consisted of twenty-five or so sheets of bark supported by a framework of sorts. This temporary housing was replaced by more permanent huts as soon as practicable. It is difficult to be specific in attempting to give a brief review of early accommodation because we are talking about settlement across a huge area of land. Nevertheless, historical records allow us to fashion a characteristic homestead and its environs.

The holding would contain one or two huts, in the latter case one for the owner and one for the men.

Lambed Down

The shades of night were falling fast
As down a steep old gully passed
A man whom you could plainly see
Had just come off a drunken spree,
Lambed down!

He'd left the station with his cheque,
And little evil did he reck;
At Ryan's pub he felt all right,
And yet he was, before next night,
Lambed down!

'Oh, stay!' old Ryan said, 'and slip
Your blanket off; and have a nip;
I'll cash your cheque and send you on.'
He stopped, and now his money's gone—
Lambed down!

He's got the shakes and thinks he sees
Blue devils lurking in the trees;
Oh, shearers! If you've any sense
Don't be on any such pretence
Lambed down.

The huts would be approximately 3 metres by 3.7 metres and cost the magnificent sum of £10. Walls were generally of slab construction while roofs were of sod, grass thatching, bark or wood shingles. In many instances men lived for years in conditions so primitive and uncivilised, even for the times, as to be called barbaric.

It is not difficult to see how this severe lifestyle eventuated. There were very few women among the settlers and so the niceties of everyday living were simply ignored. The men grew no vegetables and existed on their usual diet of tea, mutton and damper. Drinking was always a problem in the bush and without feminine influence it was even worse. Squatters had an abundance of free time while their men carried out the day's toil. Edward Curr, a squatter for ten years from 1841 onwards, relates that 'we did what we could with horses, dogs and guns to relieve the monotony of our lives, but our pastimes gradually lost their charms'.

Not far from the men's hut was constructed a primitive form of woolshed. It had enough poles to hold up a crude roof, contained a wool press and space in the centre for the sheep shearing. Sometimes this structure was built first and the men lived in it until the residence was completed.

At first, the shepherds were only able to look after relatively small numbers of sheep, the flock tallying but 200 to 300 animals per shepherd. The amount grew to between 800 and 1500 sheep depending on the terrain, timber and locale. Out-stations were built in the distant paddocks to house the shepherds and hutkeepers. These men lived lonely lives in abysmal conditions. They had hurdles, 1.8 metres long and half that in height, which were moved about to enclose the sheep at night. Forty-four hurdles could contain a flock of a thousand sheep. The hutkeeper slept in a watchbox which was also 1.8 metres long and not much larger than a phone booth. Hutkeepers cooked for the shepherds and moved the hurdles to a new site on alternate days to prevent footrot in the sheep. The animals had to be protected from both natives and native dogs, one derisive diarist of the day noting that the natives' and dogs' ideas on the rights of property were much on a par. Depending on the belligerence of local Aborigines, huts were sometimes constructed of thick timber, 5-7.5 centimetres through, with holes in the walls to allow the occupants to fire their weapons outward in defence. For all this hard yakka, the wages earned by early hutkeepers were in the region of 8 shillings per week, shepherds 10 shillings, and the big-money men, the bullock drivers, 12 shillings per week. Money was worth a lot more in those days, of course.

The source of labour during this period was commonly straight-out convicts, ticket-of-leave men and—more so with the passage of time—free settlers. Convicts were most desirable because they were of no cost at all to the landowners. If they were not submissive they earned a flogging which entailed twenty-five, fifty or seventy-five lashes, depending on the infraction. Flogging was commonplace. If it wasn't done with enough force, the flogger also was flogged. In the Hunter River region of New South Wales, property owners of the 1840s organised the employment of a full-time scourger. In some instances, large landowners had 100 or 150 convicts at their

WOOL SORTING AND CLASSING AT THE ORIGINAL BIG BURRAWONG. *NATIONAL LIBRARY OF AUSTRALIA.*

disposal at any given time. Governor Bourke set a limit of seventy convicts for one settler's use in 1835 and the *Sydney Morning Herald* promptly editorialised that such nonsense would affect 'the prosperity of the country'.

Who were the convicts? Squatter Edward Curr says: 'A heterogeneous lot they were—horse stealers, machine breakers, homicides, disorderly soldiers, drunken marines, petty thieves and so on.' What did he think of the free settlers? 'In my experience the new chums were the least satisfactory servants, though often sober men; as besides being poor hands in the bush, they were generally dissatisfied, and had a very faint idea of obeying orders. Old soldiers were better in this respect, but less intelligent than others, and generally drunkards.' Ticket-of-leave men weren't looked upon with favour by many free settlers because they were considered to be cattle and sheep duffers of the first magnitude.

So far removed from the times being discussed, it is difficult to realise the extent of gainful work conducted within the realm of the pastoral industry. In early 1846, the colony of New South Wales itself listed 13 500 of its workers as either shepherds or hutkeepers; in the outer areas fully 75 per cent of all employees were engaged in tending sheep. As late as 1891, the government census counted 556 males as shepherds in New South Wales; it is impossible to estimate the number of shepherds in South Australia because a single category contains shepherds, stockriders and station labourers, although together they amounted to 2300 males.

Right from the beginning there was a difference

THE EFFECTS OF DROUGHT ARE SEEN ON THE FACE OF THE WESTERN PLAINS. SEARCHING FOR FOOD, THE ANIMALS HAVE EATEN THE FOLIAGE FROM THE LOWER PART OF THE WILGA TREE.

in attitude among many Australians and the new chums. In *Wool Past the Winning Post*, Heather Ronald tells of the young gentlemen sent out from Britain to gain colonial experience. The servile attitude they expected at home was not to be found in Australia:

> For instance, touching of the hat was a rare courtesy from working men; the convicts had been forced to do it, so free men ever after made it a point of honour not to do so. The English lads found this 'Jack's as good as his master' attitude hard to understand, and when one pair were taken on a riding tour of Bealiba on one occasion by two of the Lee boys, they complained to Elizabeth Lee later that her sons had refused to ride behind them.

SHEEP WAITING IN A HOLDING PEN TAKE ON AN OTHER-WORDLY APPEARANCE AS THEIR EYES REFLECT THE LIGHT FROM AN ELECTRONIC FLASH.

There were attempts to create aristocracy, even by such well-known men as W. C. Wentworth. His mother was a convict and his father a medical doctor and property holder who acknowledged siring ten illegitimate children. From this background Wentworth proposed, according to author Geoffrey Dutton, 'to establish a hereditary colonial peerage'. Fortunately for all of us, Wentworth failed. Dutton adds:

> In a democratic country like Australia there is not much sympathy available for autocrats, let alone would-be aristocrats. In the radical tradition it is

WIRED FOR SECURITY.

> the convicts, the station hands, the shearers, the drovers, the bullockies, who made the Australia of the bush.

It is fair to say that today's egalitarianism is firmly rooted in the nation's colonial past.

The use of that great Australian epithet 'bloody' had become firmly ensconced in the language before the middle of the nineteenth century. Ward and Robertson in *Such Was Life* quote one Alexander Marjoribanks who wrote *Travels in New South Wales*, published in 1847:

> The word bloody is the favourite oath in that country. One man will tell you he married a bloody young wife, another, a bloody old one; a bushranger will call out, 'Stop or I'll blow your bloody brains out.'

Majoribanks kept track as one bullock driver used the word 'bloody' twenty-five times in the space of fifteen minutes.

As the pastoralists expanded into the countryside they encountered the 'blacks', Australia's first settlers. Regarding the Aborigines and the British pioneers, historian Geoffrey Blainey says that 'they were probably as far apart as any societies thrust against each other in the history of the human race'. Blainey tells the story of Hawdon and Bonney, who set out from near Melbourne in 1838, heading for Adelaide with a flock of sheep and a herd of cattle. 'For Aborigines the approaching expedition must have been unbelievable: the long-necked horses, the dust rising from the herd of cattle, the barking of Hawdon's kangaroo dogs, and the strange revolving wheels of the drays.' There were two drays, the larger of which 'carried presents for passing Aboriginals—the tomahawk was the favoured gift—and if presents should prove unpersuasive, each man carried a carbine, a pair of pistols and a bayonet'.

Friend or foe, the settlers had their way of dealing with the original Australians. And what a marked difference in the respective societies. The European believed in private property, the Aborigine believed in community property; in fact an accurate description for the then existing black culture could well be a primitive form of socialism. The European had developed agriculture and animal husbandry, while the native clung to the nomadic way of life, surviving from hand to mouth. Many of the white man's foods were able to be stored for later consumption, while Aboriginal harvests took place then and now, tribal members sharing in the day's fare. A definite ranking of individuals and groups placed varying degrees of status upon the European, while the Aborigine sprang from a system more equitable in kind. Religious belief and piety were servants to the European mind while spirits, ancestors and the Dreamtime were at the very essence of Aboriginal being. There is no way that these cultures could have met without encountering major difficulties.

From the Aborigines' viewpoint, sheep and their masters spoiled waterholes, upset wild game, disturbed their normal nomadic lifestyle and made a mockery of their sacred sites. From the other camp, colonists didn't appreciate their sheep being killed, the depredations occurring against their buildings and goods, and the frequent fires that Aborigines lit across the countryside. Not unexpectedly, these stresses led to sporadic warfare, the final outcome of which was never in doubt. Whether or not clashes took place generally depended upon the defiance of the natives. Just as with American expansion into their west, Antipodean settlers were not to be denied. If the natives were peaceable and 'retreated', there was no bloodshed; if they resisted expansion into their environment, the blacks were overcome. This guerilla warfare continued for years. An insatiable demand for new land to fuel the wool industry also fuelled ill-will between whites and blacks. A simple incident illustrates the harshness of both parties. On the Pine River, north of Brisbane, a shepherd was attacked at his hut and left for dead with a frightful skull wound which resulted from a tomahawk wielded by a native. The shepherd was found barely alive. Surprisingly, he made a complete recovery and turned down a job back at the homestead. Instead, he went back to his old hut, armed with a muzzle-loading musket. He shot and killed the first Aborigine he saw. He literally skinned the torso and then stuffed the skin with grass, suspending it from a type of gallows outside the door of his hut Understandably, no Aborigine ever came there again. Neither side shunned barbarous deeds, it was simply the overwhelming strength of the settlers that ultimately prevailed.

Shearing at Castlereagh

The bells are set a-ringing and the engine gives a toot,
There are five and thirty shearers here a-shearing for the loot.
So stir yourselves, you penners-up, what would the buyers say
In London if the wool was late this year from Castlereagh!

The man that keeps the cutters sharp is growling in his cage,
He's always in a hurry and he's always in a rage.
'You clumsy-fisted mutton-heads, you'd make a feller sick!
You class yourselves as shearers, but you were born to swing a pick.'

Squatter Edward Curr credited the Aborigines who, in their scant numbers would seem to have been most inconsequential, with results which it would be difficult to overestimate

> I refer to the fire-stick; for the blackfellow was constantly setting fire to the grass and trees, both accidentally, and systematically for hunting purposes ... we shall not, perhaps, be far from the truth if we conclude that almost every part of New Holland was swept over by a fierce fare, on an average, once in every five years. That such constant and extensive conflagrations could have occurred without something more than temporary consequences seems impossible.

Curr believed this burning-off process hardened the ground, accounted for the lack of plant mould one would expect, countered soil productivity and affected vegetation as well as insect and bird life. He concludes his theory with a sweeping statement:

> when these circumstances are weighed, it may perhaps be doubted whether any section of the human race has exercised a greater influence on the physical condition of any large portion of the globe than the wandering savages of Australia.

From what has been written here, one might conclude that the Aborigines and the men of pastoral Australia were implacable foes. Not so. In fact, without the active participation of Aboriginal people in the expansionary settlement of the continent by the European, the process would have been both delayed and more difficult. Here's how Geoffrey Dutton saw this relationship:

> The squatter needed the blacks, so instead of shooting them or chasing them off the land, many squatters saw them as friends and taught the men how to ride and manage stock. From the rich runs of Victoria's Western District to the cattle stations of the Kimberley, Aboriginal station hands have played a vital part in the history of grazing in Australia.

There were several reasons why the Aboriginal stockman was so very popular with colonial landowners, not least of which was the fact that they were a very cheap source of labour (if, indeed, they were paid at all). One immensely strong reason was that in 1851 gold was discovered in Australia. Almost overnight, gold fever infected countless numbers of pastoral workers. Men left properties for the diggings, changing forever the face and form of the pastoral industry. As one writer of the times put it, gold 'swept away the old order of things'. True enough. But these effects did not involve the black man. He didn't lust after gold and so he remained on the properties, becoming recognised as a valuable asset to the emerging wool industry.

Gold had been found as far back as two decades before the big Bathurst and Victorian finds. These were isolated, limited discoveries that had been kept quiet by those involved. Now the situation was changed and the nation was alive to gold fever. It was cataclysmic to the man on the land, for his workforce simply melted away. From shepherd to shearer, farm labour headed for the goldfields. Something had to be done, immediately. And so the nation witnessed the advent of fencing. Hundreds of kilometres of timber rail fences were put up, to be followed later by wire fencing. By the end of the 1870s, open range had passed into history. Owners discovered that sheep did quite well on their own. It was no longer necessary to hire shepherds and hutkeepers when one boundary rider could do the work of a dozen shepherds. Now the labour balance tilted in the opposite direction.

Properties existed with far fewer workers and a corresponding cost saving. There was an oversupply of men and as John Ferry says, 'a class of itinerant labourer, looking for work on the stations, became a

OPEN, WESTERN COUNTRY.

DROVERS AND THEIR DOGS MOVING A MOB OF SHEEP.

feature of the Australian bush life'. The swagman cometh.

During the early portion of the nineteenth century, the squatters were often considered to be a rough-hewn lot. Author Peter Taylor relates that this group had been described as 'money-making bachelors, half savage, half man ... half dressed, half not, unshaven, unshorn, shoes never cleaned, eating tea and damper'.

Had there been a soothsayer in 1800 predicting that the centre of the world's wool industry would be shifting from Europe to Terra Australis Incognita, he would have been laughed at or locked up. This was a society based on convicts and rough-hewn freemen sparsely situated around the fringes of what appeared to be a desolate continent, without proper materials and food to meet its own requirements, where female companionship was grossly inadequate and the land was already occupied by a hostile population. The circumstances of such a lifestyle were not likely to encourage thoughts of prosperity. 'She'll be right, mate' would not have been voiced by either prophets or settlers at the turn of the previous century.

3

Comb, CLIP *and* CUTTER

Primitive sheep do not have an ordinary coat of wool. Instead they cultivate a coat composed of hair and wool. This outer covering moults annually, the wool portion in springtime and the hair in the autumn. Allowing nature to take its course, the moulting period can last as long as a few weeks.

The earliest 'shearers' didn't actually cut away the wool but combed it from the sheep. Sometimes it could even be plucked and, as humankind advanced, a form of sickle was developed which trimmed off wool. It seems that shears were invented approximately 1000 BC in the Middle East. Within this time frame, sheep with the characteristic modern pattern of continuous fleece growth also appeared.

Shearing was both a celebration and a major undertaking in Spain early in the 1800s. About 40 000 workers looked after 5 million sheep. Working from 6 a.m. to 5 p.m. daily, shearers removed the fleeces, after which a rudimentary system of wool classing took place. Finally, a seven-step cleaning process ended with the wool being dried beneath the Spanish sun.

In Van Diemen's Land (Tasmania), sheepmen in the early 1800s only sheared their stock if the wool became too long or as a way to aid scab control. Not realising its value, pastoralists discarded the wool, leaving it to rot. They believed that the wool could be used in making a rough cloth but no-one had even bothered shipping it to textile works in Sydney.

The South Australian situation was quite different. In a Germanic settlement called Hahndorf, young women of the village took care of the shearing duties in a most unusual style. The beast was placed on its side and a piece of strong string was attached to a rear leg. The opposite end of the cord was attached to one of the young lady's toes and, when stretched out, the cord assisted her in controlling the sheep. With either the left hand or one knee, she pressed upon the animal's neck or shoulder, stabilising the sheep for the always gentle shearing process to take place. She never got through more than about thirty sheep per day, and the wool wasn't removed terribly closely, but then again the sheep were handled quite tenderly, leaving one to expect that the same animal would be returning the following year.

The general view of early colonial days was one where the property owner and his men looked after shearing duties. As the flocks grew larger and the demand for wool much stronger, it became necessary for outside help to conduct the shearing. In the beginning these shearers were often required to do sheep washing and other related jobs. However, as their skills improved they were contracted to do only shearing and became what author Patsy Adam-Smith calls 'travelling experts'. Their wages were commensurate with their abilities. In her book *The Shearers* she observes 'As early as 1826, shearers were paid 17s. 6d. a hundred, and before 1840 they were getting £1 a hundred, a price that was not eclipsed until the twentieth century'. For three-quarters of the last century shearing in Australia was done with 'blades' or hand shears. In appearance, the hand shears look not unlike the wide bladed variety of simple grass shears. Significantly, the blade shears are razor sharp. One has the impression that the shearer cuts or clips the fleece away in the same manner that scissors are used to cut hair. Nothing could be further from the truth. In fact, if one contemplates actually clipping several kilograms of wool from an adult Merino, the impossibility of shearing a hundred or more sheep a day becomes quite evident. The scalpel-like blades glide through the wool in a smooth motion, slicing off the wool. A concise description of this method was given in a journal circa 1850:

> The good blade shearer made three cuts at each blade position. With the first cut, his hand lay close to the sheep's body, so that the widest part of the blades started the cut. Then he raised his hand slightly and continued the cut with the middle part of the blades. The third cut was made with the points of the shears, with the hand raised still higher. These actions all flowed into each other with a constant turning of the wrist. While these cuts were made, the shearer had the sheep prone and his back ached with the continuous effort of coordinating several actions simultaneously.

A brief but interesting period in the development of the Australian wool industry took place between 1843 and 1851. Financially stricken graziers literally boiled down an estimated 4 million sheep in these few years. It was the onset of the 'bust' portion of one of Australia's well-known boom-and-bust cycles that panicked the growers. One Henry O'Brien, an Irish squatter, was the first to expound publicly on the scheme. Since sheep could be bought for the equivalent of 5 cents each, and sometimes even less, they were essentially valueless as they stood. Boiled down, each sheep yielded 5-7 kilograms of tallow. Sent to Britain, this was used for cheap soap and candles and even as grease for machinery. Huge vats containing up to 300 sheep converted the poor beasts into an economically viable product.

A Bagman's Toast

A little bit of sugar
And a little bit of tea
A little bit of flour you can hardly see,
And without any meat between you and me
It's a bugger of a life, by Jesus.

One squatter who owned 80 000 sheep reduced their number by 12 000 annually simply by slaughtering and cooking the animals. Boiling-down works appeared around Sydney and Melbourne, while even in the Queensland area there were eight of these plants. Geoffrey Blainey gives an account of the end products and their final destination:

> The sheep were skinned and the skins sold. The hind legs were cut off and, if near a town, were sold to butchers and salted, or smoked into mutton hams or sold as fresh meat. The skinned carcasses were then thrown into the vats and reduced by intense heat or steam. The tallow was skinned off and packed into bags made from animal hides or into wooden casks for export.
>
> Millions of hands in the British Isles were now washed with the soap made from surplus Australian livestock .

Sheep weren't put into water solely for the purpose of being boiled down—for many years they were washed prior to being shorn. This was not abandoned until the 1880s, there being three very good reasons for washing the sheep. The British market required clean wool and, true to this day, what the market wants the market usually gets. Dirty sheep often carried a burden of sand, vegetable material or burrs, depending upon where they had been grazed. The contaminants in general—and sand in particular—dulled the shearing blades and made it more likely that the sheep would be cut. Finally, the impurities added to the weight of the wool, a most significant factor since almost all of the clip was shipped to England. Increased weight means increased cost of transport.

Sheep washing was not a standard procedure throughout pastoral regions. In early times it was often a task assigned to the shearers. During the first part of the week sheep were washed and in the latter portion, beginning with the driest animals, shearing commenced. The washing process could be as simple as obliging the sheep to swim to and fro across a river or stream. As one might imagine, animals grazing in grassy, upland areas would have almost no sand contamination in their fleeces compared to those living on the arid western plains. As to the philosophy of washing sheep in cold water, where the grease would be almost completely unaffected, I can only repeat the priceless comment made by J. Garran and L. White: 'Washing sheep in cold water is only marginally more intelligent than milking a cow by having one person hold the teats, while four more lift the animal up and down.'

Another sheep-washing scheme utilised a pen or enclosure constructed of timber which would be built within a waterhole or pond-like source of water. In the basic variety, sheep were systematically passed back and forth among the workers and rubbed till passably clean. Advancements in this method saw various forms of soap used by the men in the first stages, followed by the animals swimming in clean water as a rinse. In some places, chutes or deep troughs were constructed and the labourers remained outside, directing the sheep through these passageways.

A group of Quakers in Tasmania used large tubs and hot soapy water, the men washing three or four sheep at a time. The sheep were then slid into the river where they rinsed off. In a few hours four men could put through 500 sheep. Some properties went at the job in a big way. At Brookong, they washed 5000-6000 per day. At Illilliwa, in New South Wales, forty men used hot water and a cleanser, ultimately getting through a mob of 250 000 sheep. In Victoria, a popular method was to use the water from spouts in combination with soap, the men scrubbing the animals with heated bore water.

For the men, sheep washing was not a pleasant task. Frequently they were half or nearly totally submerged with the animals, and the men had to remain in the water for hours at a time. Part of the employment covenant called for the washing labourers to receive liberal doses of spirits, in some instances a nip of rum an hour.

Another curious practice involved the use of the men's urine as a wool cleanser. Apparently, it was an efficacious lavage. Patsy Adam-Smith reports the words of a wool scourer:

> Use some of your ordinary soap ... and as much urine as you can collect in old buckets. I prefer it to ammonia. In fact, when I scoured the 'Paddington' clip, I used as much urine as the men could make, and found it thoroughly cleansed the wool, making it nice and open.

I don't know about you but I must confess to being pleased that my woollen clothing has been treated by a more modern method. In any case, when the wool industry realised that leaving the grease in the fleece improved the presentation of wool arriving at the

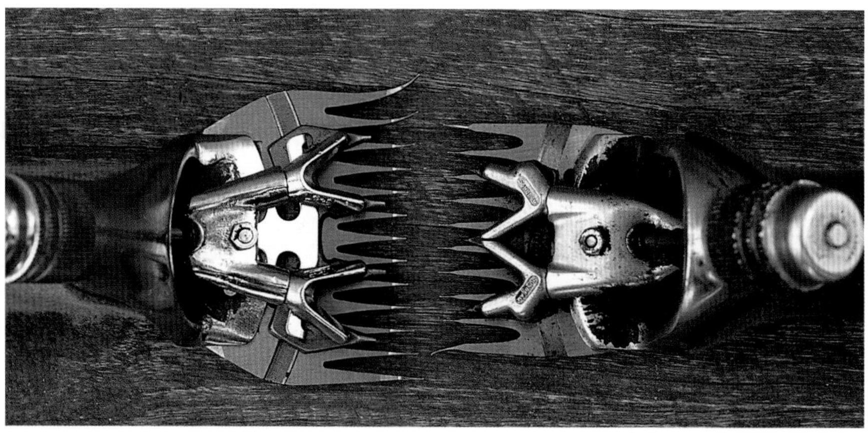

ABOVE LEFT: COMB AND CUTTER. A SHEARER SELECTS THE COMB WHICH SUITS HIS STYLE OF SHEARING AND ADAPTS TO THE TYPE OF SHEEP BEING SHORN. HE MAY ENCOUNTER DIFFERENT REQUIREMENTS NOT ONLY FROM SHED TO SHED, BUT FROM MOB TO MOB. IT IS IMPORTANT FOR CUTTERS TO BE LOADED ACCURATELY IN THE HANDPIECE SO THAT THEY DO THE BEST JOB AND MINIMISE CUTS TO THE SHEEP.

ABOVE RIGHT: COMPARISON VIEW OF WIDE AND NARROW GEAR WHICH CAUSED SO MUCH TROUBLE THROUGHOUT THE SHEARING INDUSTRY DURING THE WIDE COMB DISPUTE IN 1983.

British marketplace, sheep washing became a historic footnote. One of the few sweeping changes in the long history of sheep and shearing took place in the late 1800s. Frederick Wolseley was an Irishman, born in 1837, an immigrant to Australia at seventeen years of age. His three brothers all became soldiers, while he had a fascination for mechanical devices. Fortunately for Australia, he had an abiding interest in developing a machine which would shear sheep. Others had similar inclinations and patents had even been issued for various apparatus. The trouble was that they had all proved unreliable in one way or another.

STATION HANDS USING DRENCHING GUNS ON FRESHLY SHORN SHEEP. THE DRENCHING PROCESS HELPS PROTECT SHEEP FROM INTERNAL PARASITES.

THE COMBS, CUTTERS AND REQUISITE TOOLS THAT COMPRISE A SHEARER'S WORKING KIT.

Wolseley came up with an incomplete and imperfect device in Melbourne during 1872. In 1873, he went to Britain to study horse clippers manufactured by a local firm. Returning to Australia in 1876, he bought Euroka, a sheep station near Walgett in northern New South Wales. He continued to experiment and in 1877 he was awarded a patent for 'Improvements in apparatus for shearing sheep and clipping horses and in contrivances for driving them'. Wolseley was working with a blacksmith, Jack Gray, who was to contend that he had been the first person to successfully shear a sheep with the use of a machine. Wolseley next became involved with John Howard, who had started a small operation in Melbourne making horse clippers. A winning team was now in line. Wolseley purchased Howard's patents, employed him at Euroka and provided financial backing for Howard's investigations.

It all came together in 1885. Howard devised a fork and spring prong to lessen wear in a handpiece which is analogous to the model used today. A comparison demonstration was arranged between machine and blades at the Goldsborough and Company's Melbourne woolstore.

The match-up was between Dave Brown, a gun blade shearer, and Hassan Ali, a Khartoum native and highly competent shearer on Euroka. Six sheep were shorn, three with blades and three with the new machine. One might assume that the machine won handily. Not so. The fact is, the blades won at a canter. In the developmental days the apparatus was not all that quick; and blades had been used for decades, maximising their effectiveness. So it appeared bitter disappointment was facing Frederick Wolseley and his associates. At that moment an onlooker wondered what would happen if Ali went over the already 'splendidly shorn' sheep that Brown had done with hand shears. What happened was the beginning of the end for blades. Ali obtained a further kilogram of wool from the three sheep.

It didn't take long for the penny to drop among the pastoralists observing the demonstration. Flock sizes were considerably larger in those days and the amount of wool shorn from each sheep was much less than today's high-yielding varieties. Thus, the thought of acquiring another kilogram of wool from every three sheep made the shearing machines an instant hit among the grazing fraternity. This was not the case with shearers.

There had already been several ups and downs in the wool market and graziers could see a way to cut overheads by reducing the cost of labour. Naturally enough, shearers could see the same factors at play and they were apprehensive about their jobs. Contemporary thought held that shearing machines made skilled shearers redundant. Supposedly, it would take no more than a few days to teach the average bloke how to shear well using the new device. There was no doubt that the emerging union movement suffered a setback through the introduction of shearing apparatus. Geoffrey Blainey sums it up in a single sentence: 'The machine would thus weaken the bargaining position of the unionists by opening the work to thousands of newcomers.' Looming ahead lay a test of wills.

Dunlop was a well-known property on the Darling River owned by an empire-builder of the first magnitude, Sir Samuel McCaughey. He was a forward thinker, having installed electricity at the Toorale woolshed in 1887, the first shed so lit in Australia. It wasn't surprising that he decided to equip the forty-stand Dunlop shed with the new contrivances. When the shearers arrived and saw the installation, their reaction wasn't surprising either: they withdrew, crossing the Darling to set up a camp on the opposite shore, leaving 184 000 sheep waiting in the wings.

Wolseley's associate, John Howard, was in attendance at Dunlop and he frequently swam across the river in an attempt to convince the men to give the machines a go. It was all to no avail until, the story has it, he swam across a rising Darling twice in one morning and so impressed the men with his athletic prowess that they were influenced to begin shearing, almost three weeks after the original starting date. At first work progressed slowly, the expected problems of an early mechanical installation being compounded by sabotage to the equipment from a few of the more disgruntled shearers.

A good day's shearing with blades was 100-110 sheep. With machines the highest tally for days was less than 50, one chap reaching 75 after a week or so. John Howard again came to the rescue, offering 2 shillings and 6 pence for each sheep 'fleeced' above 150 in a single day. Once more market forces prevailed; after five weeks, three men had single-day tallies of 173, 168 and 167. It wasn't all that long before each member of a seven-man team shore 200 sheep in a single day. The historical record of first having the entire clip cut by machine shears belongs to Dunlop. By the end of 1888, eighteen properties had shorn by machine and their use spread further in 1889.

What happened to our two principals? Wolseley sold Euroka in 1889, travelled to England and set up the Wolseley Sheep Shearing Machine Company in Birmingham. In 1893 he had the chief engineer of his Sydney company, Herbert Austin, come to Britain to manage the local factory. Wolseley was in poor health so he left the company in 1894. The next year Austin created a new automobile named after—guess who?–Wolseley. The latter died in London early in 1899.

Howard travelled the Antipodean countryside with a distinguished shearer, Jim Davidson, demonstrating and selling the machine shears. Howard worked for Dangar, Gedye and Malloch Ltd, who merchandised machinery and he remained in Australia until he died in 1920.

Machine shearing transformed the wool industry. Quite simply, it got more wool off more sheep. Let's examine some practical aspects of both methods. Incidentally, some blade shearing is conducted even today. It is done in isolated parts of Europe and the Americas, in some small British flocks and more so in South Africa. In Australia there are still a few men alive who can wield the blades. Nick Trompf interviewed 79-year-old Frank Morris of Rokewood, Victoria, who estimated that there are currently 'no more than ten blade shearers left in Victoria'. Frank's best tally was 111, not a large figure compared to some of his colleagues' numbers. 'That's why I've lasted a bit longer, by not taking it out of myself on the board.'

Duke Tritton (1886–1965) was a Sydneysider who 'tramped around the bush for four years, from 1905 to 1909', and later wrote what has become an Australian classic, *Time Means Tucker*. He did fencing, trapping, droving and other rural jobs but mostly he was a shearer. He was at a property, Cowimbia, to observe its final season of blade shearing. Here's how he described it:

> After the noise of the machines it was very quiet. Contrary to the general opinion and the well-

known song, shears do not click.

There seemed to be more rhythm in a 'blade shed', possibly because of the lack of noise. A big 'machine shed' sounds like 10 000 locusts on a hot day, with the whirring of the machines and the hum of the overhead gear and friction wheels. The only sound in a blade shed is the 'chop, chop' of the knockers and a low ringing noise as the blades meet. As an onlooker I found it very soothing.

The hand shears were usually altered by attachments the men put on to make them work more effectively. As they were the main tool in their craft, long hours were spent in getting the shears just right. Duke Tritton again:

> The gullets of the hand grips are filled with soft wood or sometimes cork. This stops the heels of the blades from meeting, so the sound is a soft 'chop, chop'. These are known as 'knockers'. Then there is a band of leather from the heel of one blade to the back of the other. This is the 'driver', and its purpose is to prevent the hand slipping forward on the blades. Then there is the 'stopper', another strap which prevents the blades from opening to their full extent.

The topic of sound in the woolshed did not foster unanimity among shearers. Blade men disliked the noisy machinery, claiming it aggravated their nerves. Machine men could not bear the quiet, which was boring and tedious to them.

A major difference in the two methods is the way in which the sheep are handled. Godfrey Bowen explains:

> A good blade shearer works around his sheep and can shear it in almost any position on the board. If it should struggle or kick one way or the other this does not worry a blade man.
>
> It is not so with machines. The machine will not come off the wall and follow you round the board: the sheep must be in the correct position for machine shearing. This is why machine shearing is harder, or takes longer to learn than blade shearing, but when it is mastered it is probably much easier.

Sharpening the cutters on machine shears is a simple, quick chore whereas it has been described as tedious for hand shears. The blade shearer has to stop more times during a run to hone his cutting edges. Machine shears are quicker to cut sheep but the razor-sharp blades of hand shears do more damage when they do slip; stitching is required more commonly with blades.

Which method is faster? In a word, machines. There were big tallies accomplished by blades but the sheep were different in those bygone days. They had neither the amount of wool nor trimmings that are found on modern sheep. Hand shears are drastically slowed by the leg, head and crutch area trimmings.

Everyone knows that shearers often assumed the role of swagmen and trudged over the countryside seeking food and fortune. Less well known is the importance attached to ordinary bicycles in transporting a mobile labour force across the backblocks of the nation. In the 1890s, bicycles became the shearers' Holdens, a position they retained through World War I and in some places right up to the 1930s.

Tasmanian shearers left their island home in March, bicycles at hand, and caught the boat to Melbourne. From there they would take a train as far as practicable and start out on their bicycles, covering ground that even today kills unwary strangers.

Bicycles possessed definite advantages over horses: they didn't have to be hobbled to graze or be caught and saddled before beginning the day's journey. Neither did they come down with equine viruses or suffer from other illnesses. All this contributed to a light-travelling shearer who covered great slabs of countryside. Men commonly rode 2000 kilometres in a season and individuals were known to wend their way across 5000 kilometres of back country in a single year.

Everyone travelled lightly: a quart pot, a change of clothes, a blanket, perhaps an oilskin for a groundsheet and—a compulsory item—a waterbag. C. E. W. Bean wrote about this aspect of bicycle peregrination in 1910:

> The shearer set out on these trips exactly as if he was going from Sydney to Parramatta. He asked the way, lit his pipe, put his leg over his bicycle,

SHEEP ARE 'DIPPED' AFTER SHEARING AS A DEFENCE AGAINST EXTERNAL PARASITES SUCH AS LICE, TICKS AND ITCH MITES. HERE THEY ARE BEING PACKED INTO THE RACE PRIOR TO ENTERING THE SHOWER DIP ITSELF. LIKE CHILDREN IN A BATH, SHEEP ARE NOT OVERLY KEEN ON BEING DIPPED. IT IS ALL OVER IN A MOMENT AND THEY EAGERLY AWAIT THE OPENING THAT LEADS TO FREEDOM.

A DRY RIVERBED PROVIDES MUTE TESTIMONY TO THE EFFECTS OF INLAND DROUGHT.

ABOVE LEFT: SHEARING SHED ACTIVITY AMONG SHEARERS AND ROUSEABOUTS.

ABOVE RIGHT: THE 'EXPERT' GENERATES SPARKS SHARPENING COMBS AND CUTTERS. THEY MUST BE GROUND WITH HIGH RPMS TO 'BRING THEM UP QUICKLY' AND PROVIDE A GOOD CUTTING EDGE.

and shoved off. For precisely the same trip the average European would probably requisition a whole colonial outfit, compasses, packhorses, pugaree, sun-spectacles, and field-glass. The native Australian took it like a ride in the park. If he was city-bred, as were many shearers, the chances were that he started in a black coat and bowler hat, exactly as if he were going to tea at his aunt's.

Punctures were commonplace. Today's immobilising mishap was not more than a minor inconvenience to yesterday's cycling shearer. When the tubes became too ragged to patch they were filled with rope, bits of clothing, even grass. On the outside the tyres would be affixed to the rims with bindings of kangaroo hide, canvas or worn trousers. It seemed that wherever man could tramp, the bicycle could accompany him. The tyre tracks of bicycles were found in the furthest regions of the bush, from saltpans to desert.

Turning to the shearers themselves, the pseudonymous Rolf Boldrewood was a colonial author who wrote *Shearing in the Riverina 1865*. He provides a well-drawn portrait of the men who forged the wool industry more than 130 years ago. Boldrewood's picture is an evocative one:

Seventy men, chiefly in their prime, the flower of the working-men of the colony, they are variously gathered. England, Ireland, and Scotland are represented in the proportion of one-half of the number; the other half is composed of native-born Australians. Among these last—of pure Anglo-Saxon or Anglo-Celtic descent—are to be seen some of the finest men, physically considered, the race is capable of producing. Taller than their British-born brethren, with softer voices and more regular features, they inherit the powerful frames and unequalled muscular development of the breed. Leading lives chiefly devoted to agricultural labour, they enjoy larger intervals of leisure than is permissible to the labouring classes of Europe. The climate is mild, and favourable to health. They have been accustomed from childhood to abundance of the best food; opportunities of intercolonial travel are frequent and common. Hence the Anglo-Australian labourer without, on the one hand, the sharpened eagerness which marks his trans-atlantic cousin, has yet an air of independence and intelligence, combined with a natural grace of movement, unknown to the peasantry of Britain.

4

RIVALS

Along with the sheep, the paddocks of Australia contain three other highly visible animals. These 'beasts of the field' are the rabbit, the dingo and the kangaroo. All three have affected the health and the numbers of sheep to a varying degree.

Rabbits

The amazingly fecund Australian bush rabbits are fascinating animals. Their powers of reproduction, their ubiquitous distribution which seemingly occurred in a trice, the diverse methods employed towards eradicating them and, finally, the unbelievable boondoggling that went on when the answer to zapping the rabbit had been at hand for thirty years make for an intriguing story. Here then is an account of the rabbit, the diminutive creature that has made a bunny out of all of us and cost the sheep-wool industry not millions but billions of dollars over the last century-plus.

When and where did this environmental nightmare begin? In beautiful downtown Sydney where the nation's early sheep got their start? No. It was close to Melbourne, Sydney's friendly rival in almost every category. It was begun by a rich Pommy who, like many of his fellow countrymen, wanted to re-create a wee bit of England in the colonies. Thomas Austin, the squire of Barwon Park, a prestigious property near Geelong, received two dozen rabbits from the clipper ship *Lightning* shortly after the Christmas of 1859. These original twenty-four were British cousins of the European rabbit, *Oryctolagus cuniculus cuniculus*.

As well, seventy-two partridges and five hares had been ordered by the squire so he and his friends could enjoy a jolly good bit of sport.

Once released there was no stopping the rabbits, which isn't surprising when one considers their breeding capabilities. It is estimated that a lone pair of rabbits can yield in the neighbourhood of 9 million offspring in the space of three years! The doe can give birth at four months and, if all goes well, she can repeat the process as many as six times a year, producing from one to eleven kittens—young rabbits—each time around. The figures are truly staggering. Depending on the authority one accepts, six to ten rabbits eat the same amount of pasturage as does a single sheep. Thus, a property supporting 6000-10 000 rabbits has to carry 1000 fewer sheep. While this may not be a problem during good seasons, drought periods render the food situation crucial.

Yes, rabbits eat the usual grasses that sheep feed on. But it doesn't stop there. After destroying pastures they eat seedlings and ringbark trees. As Ivan Letchford put it, 'They even eat the bark off the scrub growth and that's what keeps sheep alive in times of drought'. Widespread overgrazing is a precursor to erosion, both wind and water becoming attacking forces.

WOOL PRESSING HAS ALWAYS BEEN HARD WORK BUT TODAY'S TECHNOLOGY EASES THE BURDEN. *NATIONAL LIBRARY OF AUSTRALIA.*

Returning to Thomas Austin, we find that his rabbits have taken hold with a vengeance. The *Geelong Chronicle* of 2 May 1862 reported that the imports destroyed the young trees which had been planted as cover for pheasants and partridge. By July 1865, this time in the *Geelong Advertiser*, Mr Austin was reported to have killed 200 000 rabbits and he had but 10 000 remaining. Obviously Mr Austin did not eliminate the remaining animals. Instead, they spread like wildfire across Australia.

Reaching South Australia and New South Wales within twenty years, they continued on to arrive at the Queensland border after travelling 800 kilometres in seven years. The broad Nullarbor Plain proved to be a poor defence and by 1894 rabbits were seen in Western Australia. It can be said that rabbits occupied the southern rather than the northern portion of the

IT IS IMPORTANT THAT FLEECES BE PROPERLY THROWN-OUT AND SPREAD FLAT SO THAT THEY CAN BE SKIRTED RAPIDLY BY THE WOOL-ROLLERS.

A MODEST TEAM, CONTAINING SIX SHEARERS, WILL GENERATE 800 OR SO FLEECES PER DAY, KEEPING EVERYONE MOVING BRISKLY.

continent, the Tropic of Capricorn acting as an approximate northern border for their advance. They weren't spread evenly over the land, but were concentrated in areas which the rabbit found favourable. 'Rabbits aren't silly', says Ivan. 'They prefer the sandy, light country to the heavy black country. They don't burrow well in black ground. Besides, it tends to fall in on them when it rains.'

The number of rabbits involved almost defies imagination. In 1869, William Robertson believed he had killed nearly 2.1 million on his property. Thomas Guthrie of Brim station, along the Wimmera River, suffered a rabbit plague in the late 1870s. When they mustered the sheep there were 670 live animals where there had been 120 000 the previous year. The rabbits had eaten all the feed that normally went to the sheep. Mr J. Armstrong of Gunbar station had 360 000 acres (146 000 hectares). It was his estimate that there were 100 rabbits on each acre. If his estimate was accurate, then his property was overrun by 36 million rabbits. Eric Rolls wrote, 'In the first eight months of 1887 there were 10 million rabbits destroyed in New South Wales.' Some properties were killing 10 000 rabbits each week.

The famous Falkiner family of Boonoke spent a small fortune trying to eradicate rabbits. In 1894, they spent the very large sum of £12 000 on their endeavour. Tim Hewat quotes Otway Falkiner regarding his younger years: 'One of the worst jobs I had to do in those days was to clean out the rabbit bays along the fences. Dead and stinking rabbits were waist high around me as I worked.'

Some individual landowners fenced their land in an attempt to keep rabbits off their properties, a procedure that met with mixed success. Other measures included poisoning, fumigation, trapping, shooting, ripping and dogging. It sounds like all-out war. However, even with the battle being joined on several fronts, some of the foot soldiers were shamming.

In an attempt to gain the upper hand, poison carts were tried in New South Wales. Several carts were used on a property. Spaced thirty metres apart, the carts dug a small furrow into which the poison phosphorus bait was deposited. This machinery was dragged across several kilometres of paddock before the drums of poisoned bait ran out. The result was temporary at best, the villains replenishing their numbers almost overnight, or so it seemed. Besides, the poison killed a considerable number of birds.

Duke Tritton was a shearer who wrote of his experiences in the bush. He tells about digging pits with trapdoor covers alongside rabbit fences:

> As the rabbits were falling in day and night and piling up, the lower ones smothered and the pit

would be a mass of putrid bodies. It was the pit-runner's job to clear these foul smelling places.

Apart from stinking rabbits there were other unpleasant things which made the pits unpopular. In the summer snakes and goannas would often be found in the pits. To slide down a hole and find a bad-tempered snake eager for a fight is an experience not easily forgotten. Goannas were never very friendly, either, after being shut up for a few days.

Boundary rider Lester Teague remembers Upper Hunter areas being overrun with rabbits. At one time over a thousand hectares were ripped with bulldozers, the blades exposing the rabbit warrens. Dogs got those that weren't destroyed in the initial onslaught. Pastoralist Ron Taylor recalls a gang of sixteen men, bearing a hundred traps each, working 1600 hectares in the Upper Hunter. He personally counted 40 000 skins during one active period. In the same region, Ivan recollects rabbiteers going to the pound at Newcastle to collect mongrel dogs; they maintained packs of 50 to 200 to be used in rabbit control. The poison sodium fluoracetate, known commercially as 1080, is generally spoken of as the most successful of chemical measures. If used when the rabbits weren't breeding madly (autumn), it could attain a 99 per cent kill rate.

Rabbiteers were notorious for perpetuating the rabbit plague. After all, their livelihood depended on it. The skins were of little value but the carcasses were a different matter: 1.5 tonnes were auctioned daily at the Melbourne Fish Market. A movement began which favoured exporting rabbit carcasses. This is Eric Rolls' wry comment: 'Encouraging this trade because it was good for rabbiteers was like encouraging the smallpox because it was good for doctors.' It went ahead anyway.

And so we come to an amazing story, a story of bureaucratic bungling well beyond officialdom's usual incompetence. It began in 1919 when a South American correspondent informed the Australian government that rabbit myxoma, a virus specifically acting upon the European rabbit, had been 90-100 per cent effective in South America. The Commonwealth Institute of Science and Industry rejected myxoma out of hand because Australian 'popular sentiment' was against the use of a disease virus. A sample did arrive in 1926, only to fail in tests because the locals didn't know it required an insect vector to transmit the virus to the rabbits.

Enter the hero or, in this instance, the heroine. Dr Jean Macnamara, a Melbourne physician studying in America, learned that myxoma was destroying rabbits in California. She arranged for the virus to be sent to Australia in 1933. Don't cheer yet: the Director General of Health had the virus destroyed. Dr Macnamara persisted. In 1936, the Council for Scientific and Industrial Research (which later became the Commonwealth Scientific and Industrial Research Organisation [CSIRO]) began Australian trials. Done inadequately, results were poor and testing was abandoned in 1943. Dr Macnamara became politically active and managed to obtain fresh trials in 1950. Again because of poor methodology, results weren't encouraging. The CSIRO said in August 1950 that myxoma had failed anew. Fortunately, trials continued and this time around the myxoma took hold with a vengeance. By 1951, over an area of 2.5 million square kilometres, innumerable outbreaks of myxomatosis were reducing the rabbit population by 80-100 per cent, depending on locale and circumstances. It was estimated that 200 million rabbits had been destroyed by 1953. Of course, that led to the good news: Australia's wool clip jumped by approximately 32 million kilograms, directly or indirectly increasing prosperity for everyone.

Myxomatosis is a cruel disease. The affected animal develops a discharge in its eyes and the lids may even become closed over with yellowish matter. Soft lumpy tissue swellings occur any place on the rabbit's body and the end comes in less than two weeks. The disease is characterised by summer outbreaks and this coincides with increased activity by the vector transmitting the virus. Some rabbits do recover and this leads to a gradually increasing level of resistance.

The rabbit problem had not ended, but the outlook had immeasurably improved. Without Dr Macnamara's dedicated persistence, who knows how long the bureaucrats would have squabbled, frittering away Australia's economic prosperity?

On a lighter note, the rabbit story did have its absurdities. One W. E. Abbott was president of the Central Stock Board of Advice in New South Wales in 1913. Because cats ate rabbits, Mr Abbott devised a scheme to employ the former to get rid of the latter. He even published his scheme in a booklet to give rural folk an opportunity to study his figures.

Starting with 100 000 acres (about 40 000 hectares) of land populated by 10 000 rabbits, add 100 cats. It's over to Eric Rolls (*They All Ran Wild*) for quotation:

> *1st Year*
> Rabbits would increase to 60 000 at end of first year. Cats would require for food in first year 36 500 rabbits. Total number of rabbits at end of first year 23 500. Cats have now increased to 300.
> *2nd Year*
> Rabbits would increase to 141 000 at end of second year. Cats would require for food in second year 109 500 rabbits. Total number of rabbits existing at end of second year 31 500. Cats have now increased to 900.
> *3rd Year*
> Rabbits would increase to 189 000 at end of third year. Cats would require for food in third year 328 500 rabbits.

Using Abbott's figures, Eric Rolls calculated that 1125 cats would get the last rabbit before March of Year Three. Mr Abbott doesn't deign to mention what happens to the 1125 now-feral cats, hungrily prowling the bush looking for extinct rabbits.

There is a current chapter to the story of European rabbits in Australia. Scientists estimate their numbers at 300 million in 1995 and government spokespersons believe they eat $90 million worth of crops each year. So it would seem to be a wonderful idea to rid Australia of this European import. Research was being conducted on Wardang Island, a few kilomentres off the South Australian coast, involving the rabbit calicivirus as a control agent for rabbits just as myxomatosis was used in the past.

The calicivirus was first seen in China in 1984 and is now in several countries. The virus causes a haemorrhagic disease in rabbits whereby they die of heat and lung failure within forty hours of infection. The disease travels at fifteen kilometres a month in Europe but in Australia it sprints over the countryside at up to five kilometres per day, killing 80-90 per cent of the rabbits within six to eight weeks of its arrival in an area.

Another Fall of Rain

Now the weather had been sultry for a fortnight's time or more,
The shearers had been battling might and main,
And some had got the century as never had before
But now all hands are waiting for the rain.

Chorus:
For the boss is getting rusty, and the ringer's caving in
His bandaged wrist is aching with the pain
And the second man I swear, will make it hot for him,
Unless we have another fall of rain.

A few had taken quarters and were coiling in their bunks
When we shore the six-tooth wethers from the plain.
And if the sheep get harder, then a few more men will funk,
Unless we get another fall of rain.

But the sky is clouding over and the thunder's muttering loud,
And the clouds are driving eastwards o'er the plain,
And I see the lightning flashing from the edge of yon black cloud,
And I hear the gentle patter of the rain.

So, lads, put on your stoppers, and let us to the huts
Where we'll gather round and have a friendly game.
While some are playing music and some play ante-up
And some are gazing outwards at the rain.

But now the rain is over, let the pressers spin the screws,
Let the teamsters back the wagons in again,
And we'll block the classer's table by the way we push them through,
For everything is merry since the rain.

And the boss he won't be rusty when his sheep they are all shorn,
And the ringer's wrist won't ache much with the pain
Of pocketing his season's cheque for fifty pounds or more,
And the second man will ride him hard again.

Words based on a poem by John Heilson

Now the question is, how did the virus escape from Wardang Island to race across the Australian mainland? There are no certainties but a favoured theory holds that flying insects such as bushflies and mosquitoes, carried by strong winds, could easily have transported the disease to the continental landmass. Regardless, now the disease is travelling like a bushfire, scientists for the program have released the virus at nearly 200 sites with widely successful results.

All that's left to do is to deal with the unexpected effects of this biological weapon. Akubra, makers of fine hats from rabbit fur, have had to import skins from Europe and New Zealand. Specialist food suppliers of wild rabbit meat for both overseas and domestic markets are suing government agencies for the loss of their livelihood. Outback rabbit shooters are out of business and must seek other employment. Asian, American and European importers will not accept Aussie rabbit meat unless it is guaranteed free of the virus. So the humble rabbit continues as a controversial subject in its adopted home and it's safe to say that we have yet to hear the last word from this resilient animal.

ABOVE TOP: SKIRTING A FLEECE REMOVES BURRS, SEEDS AND FOREIGN MATTER PICKED UP BY SHEEP GRAZING IN OPEN RANGE CONDITIONS.

ABOVE: THE SKIRTED WOOL IS OF LESSER COMMERCIAL VALUE THAN THE MAIN FLEECE WOOL.

Dingoes

The Australian native dog has become a bone of contention between pastoralists and environmentalists. As expected, the former would like to eradicate dingoes while the latter hope to preserve the animals. There is a story which looks at this dispute in a humorous light. However, it seems to fit into the category of a yarn rather than a factual report In other words, it's probably a tall story.

Just a few years ago a meeting was held in the Newcastle area. This get-together was an attempt to work out some ground rules for managing the dingo problem. There were graziers, contractors, animal lovers and, no doubt, other interested parties.

Various individuals were airing their views on how to cope with dingoes. Naturally enough, some folks thought that a bullet between the eyes would be the best solution. Just as naturally, others believed that the dingo must be preserved at all costs. After many speakers aired their views, a speaker with 'three strikes against him' took the stand. The three strikes? The speaker was a woman, a university graduate and an environmentalist. She began to speak, focusing on castration as the favoured method of controlling dingoes. It was her plan to trap the dingoes humanely, castrate the males and release them, allowing the dogs to run free. She felt that such a program would preserve these beautiful animals and keep them from becoming too numerous.

After a few moments, a rough-hewn man in the audience stood up and interrupted the speaker. 'Madam,' he said, 'you don't seem to understand the problem at all. You see,' he continued, 'the dingoes are eating our sheep, they're not fucking them.' So the story goes!

The dingo looks enough like man's best friend to be mistaken for him at a distance. A dingo of average height would be 0.75 metre tall and 1.5 metres long from tip of nose to tip of tail. The tail is thick and bushy, ears are upright and the coat is coarse-haired. The dingo does not bark but produces a kind of howl. The animal attacks and kills sheep, of that there is no doubt. Whether it attacks and kills humankind is an agonising, complicated and unresolved question.

In a recent review, Roland Breckwoldt found that dingoes did not travel widely. In the New England Tablelands, each adult utilised approximately thirty square kilometres of territory. The standard diet comprises 'native fauna and rabbits' although it is widely accepted that they kill well in excess of their food requirements. If a great number of sheep are killed wantonly, one reason is that a dingo bitch is giving her pups a lesson on how to hunt. The usual method of dingo attack is for the animal to charge into its prey, striking with strong jaws and leaping back again.

Dingoes arrived in Australia a few thousand years ago, either by land bridge or by sea, the latter explanation perhaps finding greater acceptance. They were camp followers of sorts, hanging around the natives' settlements. Dingoes were not regarded with disfavour by the earliest Australians because it is known that native women fed the pups by placing them to their breasts. During cold nights the mature

CLASSING WOOL. BOTTOM: THE TRADITIONAL METHOD OF CLASSING ASSESSES THE SPINNING QUALITY—FINE, MEDIUM OR STRONG —OF EACH FLEECE.

TOP: THE CLASSER EXTENDS A STAPLE OF WOOL FROM THE FLEECE TO APPRAISE ITS VALUE.

animals served as living blankets for chilled sleepers.

No sooner had colonists begun grazing sheep than dingoes began their depredations. Shepherds were necessary to protect those early flocks, particularly during the hours of darkness.

Individual 'rogue' dingoes have been known to kill a hundred sheep in a single night. Duke Tritton was told by a dog trapper (often called a dogger) that 'pack dingoes, apart from an occasional lamb, did little damage to stock, being content to live on rabbits and the native animals'. On the other hand, the loner 'would get into a flock and kill for the sake of killing. The pack would feed on the dead sheep and were blamed for the slaughter.' Tritton knew of a dingo who had apparently slaughtered over a thousand sheep; when he was finally done-in, the bounty paid was £56, a considerable amount of money early in the century.

The bitter poison, strychnine, was introduced in the 1840s. As it proved effective in the elimination of dingoes, fewer shepherds became necessary to watch over the flocks. Western Victoria was the first area cleared of dingoes, in the 1850s, from which time shepherds were dispensed with.

Doggers were a strange breed of men who made good money if they were adept at their job. A bounty paid by the Pastures Protection Board could be supplemented by a payment from one or more landowners. From 1880 to 1902, the New South Wales Pastures and Stock Protection Board paid for 15 174 dingo scalps. A scalp consisted of two ears and a forehead. There is a tale concerning a Chinese dogger who was hired on one property only to disappear into the bush. He began to bring back numerous scalps, much to everyone's surprise. As C. E. W. Bean explains:

> Nobody knew there were so many dogs about; nobody else saw them. But the Chinaman was bringing in whole mobs of scalps. In the end someone examined the scalps. They found that when he killed one dog he manufactured about twenty scalps out of him—ears and all. They were fine art. But after that they altered the rules. Since that date a scalp has meant a strip of skin cut from the tip of the nose right down the back to the tail. You cannot manufacture two backbones out of one dog.

A trapped dingo goes wild when the steel jaws bite. That's why some of the doggers preferred poison; the ruckus set up by the trapped animal would scare others away if they were in a group. One Queensland trapper was known to place a mixture of cloth rags and strychnine crystals on the jaws of a trap. When caught the dingo would bite at the rags around its lower leg and poison itself.

The matter of poison for dingoes is a contentious issue. Today's poison of choice is 1080, mentioned previously in regard to rabbit control. It is toxic to other wildlife and 1080 detractors point out that meat-eating creatures such as possums, bats, bandicoots, lizards and birds are at peril from exposure to the substance. Longtime grazier Ron Taylor does not address the problem of wildlife damage but he does say that 'in all the years 1080 has been used no-one has been accidentally poisoned'. Ron speaks with personal conviction about the dingo issue:

> I was a member of the Dingo Board in Scone for just on twenty-eight years. The Dingo Destruction Board they called it for a start. Then they got it to the Dingo Control Board, trying to placate these do-gooders, and now each district has a Land Protection Board. It was as if we were doing something cruel to these animals. Never mind what these dingoes are doing to other animals. Then they got to the silly stage where because of our control of the dingo we were upsetting the ecology and the balance of nature. I asked them: Over in New Zealand they've got no black snakes. Should we export some to them? Over in Ireland they've got no blowflies. Should we export some to them? A lot of this ecology business is silly rot.

It is Ron's opinion that the dingo population has gone down where aerial baiting has been utilised, but elsewhere their numbers have increased. He identifies two factors as significant in this increase: the severe restrictions on baiting and the decreased strengths of modern baits.

One defence measure against the dingo has given Australia a record of sorts. The dingo fence which runs in parts of Queensland, New South Wales and South Australia is about 9600 kilometres long, reported to be the longest man-made structure in the world. Boundary riders are permanently employed to ensure the integrity of the fence. Does it work? There was no general agreement among respondents but there was one rural wag with an entertaining comment: 'Listen, mate, without that fence the dingoes would be in North Sydney.'

Kangaroos

Any discussion of kangaroos is sure to put the metaphorical fox among the chickens in so far as pastoralists and environmentalists are concerned. Does the roo expropriate food better directed to sheep? Yes and, probably, no.

Kangaroos and wallabies are all macropods, a group embracing forty-five species. This discussion deals only with the grey and the red roo. The grey is smaller, a resident of timbered and scrub habitat; the larger red prefers open country. A mature red may measure over 2.1 metres in length, weigh over 45 kilograms, maintain a speed of 40 kilometres per hour while leaping 1.8–2.4 metres in the air via bounds which are 7.5–9 metres in length.

Kangaroos are in no hurry to conclude the mating game, which continues for as long as twenty to fifty minutes, depending on the category of roo. (Perhaps this explains the slight grin scientists have detected in the specified animals.) There is no particular season for reproduction. In the case of a red female, she will breed first between fifteen and twenty-four months of age, after a 33-day pregnancy giving birth to an inconspicuous nubbin of flesh weighing about one gram.

The tiny joey squirms its way over the mother's body, reaching her pouch in just a few moments. Attaching itself to one of the mother's teats, it will receive sustenance from a thin watery liquid. Later, the joey is nourished by milk which is increasingly heavy and creamy. Not until half a year later does the joey actually leave its mother's pouch for good.

Observers have stated that kangaroo numbers seem to have increased quite notably attendant upon the coming of white settlers. There are two main reasons given to explain this. The first concerns new sources of water necessary for sheep grazing which are likewise available to the kangaroo. Secondly, sheep grazing habits leave the food that roos thrive upon, encouraging their expansion.

Another line of thought holds that kangaroos increased as dingoes were eradicated and the Aborigines were overwhelmed. Dingoes successfully nabbed the young joeys while Aborigines culled the adult roos.

Yet another voice casts doubt on the preceding analysis. This reasoning claims that close scrutiny of journals kept by early explorers suggests that there is little difference in the overall kangaroo population from early colonial days until now. About the only statement beyond dispute is that there are no fewer roos today than there were prior to the arrival of white settlers.

Graziers certainly noticed increased roo numbers. In 1851, one landowner's wife developed a fondness for several roos on the property, her attachment guaranteeing their safety. During a three-month period in 1875, there were 20 000 kangaroos shot on this station. A property owner near Geelong killed 3000 in a single day. At Outalpa, in South Australia, 14 000 were destroyed in two years while at the neighbouring Oulnina, 16 000 were put to death. At Gordon Downs in Queensland, 61 000 were killed in a twenty-month period.

Pastoralists contend that kangaroos destroy fences and devour food which is intended for stock animals. That's probably true, to a degree. During times of normal pasture growth, there are no problems and there are even studies which indicate that sheep and roos consume different grass. It's the rivalry for herbage during droughts that causes hostility among graziers. The landowners' position is well stated by Eric Rolls when he notes that 'graziers bitterly denounced the conception that they should maintain a public zoo at their own expense'.

Environmentalists do not always see eye to eye with pastoralists. The former decree that cloven-hoofed sheep compact soil and growth while cushioned kangaroo paws are easy on the land and herbage. Further, roos ignore paddocks with grasses they dislike, allowing areas to lie fallow for months at a time. Unlike sheep, roos in drought situations have a self-regulating mechanism to control births, thus preventing native grasses from being overgrazed to the point of destruction. The ecologically minded are accurate when they claim that sheep are hard on the fragile Australian environment. Sheep require four times more water than roos, their metabolic rate is

KANGAROOS OFTEN COMPETE WITH SHEEP FOR AVAILABLE FEED. THEY ARE A FREQUENT TOPIC OF DISCUSSION IN SHED ENVIRONS.

higher and 'heat stress' causes more problems for sheep. All this has led some people to suggest that in the vast western areas where sheep production is marginal at best, a feasible alternative would be to inaugurate kangaroo farming.

On the face of it, the proposal seems to have merit. Roos have approximately twice as much 'carcass muscle' as do sheep. As an example, a sheep weighing 35 kilograms yields 9 kilograms of boneless meat. A roo weighing the same would yield 18 kilograms of boneless meat, quite a surprising difference. Roos, on the other hand, are not as efficient in managing their food intake. 'Although kangaroos have a digestive system similar to ruminants,' explains Roland Breckwoldt, 'they are less able to digest low protein fibre than sheep and cattle.'

What's preventing the adoption of roo farming? In a word, uncertainties. Australian wildlife belongs to the Crown, not to the private sector. Who has final

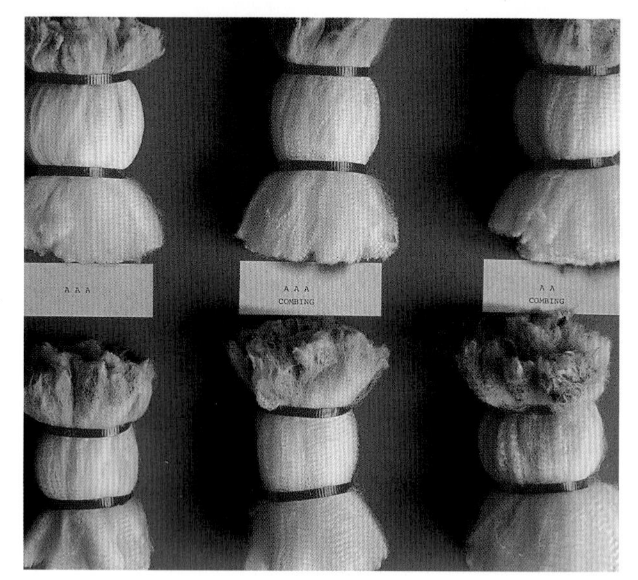

CLUSTERS OF STAPLES ARE FROM FLEECES SELECTED RANDOMLY AND SHORN FROM AUSTRALIAN MERINOS CARRYING TWELVE MONTHS' GROWTH. STAPLE TIPS SHOW EFFECT OF WEATHER THROUGH THE GROWING PERIOD. THE TOP ROW IS FROM A TYPICAL MEDIUM CLIP GROWN IN THE NORTH-WEST DISTRICT OF NEW SOUTH WALES. THE BOTTOM ROW IS STRONG WOOL FROM THE WALGETT AREA. (THE WOOL IS, LEFT TO RIGHT, GRADED AAA, AAA COMBING AND AA COMBING.)

THE PLEASURES OF BUSH LIFE INCLUDE VISUAL BLESSINGS FROM NATURE.

PLENTIFUL BIRD LIFE ENHANCES RURAL REGIONS.

control? There's no guarantee of profitability in the industry. What if nations ban kangaroo imports? Roo management is another stumbling block. Won't the costs of high-quality fencing to contain roos inhibit returns? All of these uncertainties combine to create irresolution. Besides, with the price of wool as it stands today, even a marginal property owner catches glimpses of a pot of gold at the rainbow's end.

Rabbits, roos and dingoes have confronted sheep and graziers extending well beyond the past century. The balance of power has changed numerous times. It would be worthwhile if Australians of integrity could agree on a policy which would yield sound economic benefit to the nation while preserving its wildlife heritage. Surely an equitable balance can be attained.

In concluding this chapter, a cautionary note for urban readers and overseas visitors. Kangaroos are nocturnal animals and they can severely damage a modern automobile being driven at speed during the hours of darkness.

Roos feed adjacent to the roadway, especially during drought conditions. Run-off moisture from the paved surface encourages plant growth which in turn attracts grazing wildlife. Drivers have counted scores of kangaroos at road's edge during a single night's journey in drought periods, when usual sources of food were unavailable.

There is a certain brand of rural logic which declares that it is best to drive at high speed during night hours and this way the vehicle will be past the roo before it has a chance to jump out on the highway. This advice should be ignored.

Drivers frequenting country roads at night are advised to beware of 'overdriving' their headlights. It is easier to prevent roo damage than it is to repair it.

Lester Teague, Boundary Rider

I was born in Moonan and spent all my life around here. I've been to Sydney maybe ten times, to Tamworth twice, to Newcastle and Singleton. I've never been to Armidale or even Taree. I've never been out of the state but one day I did go to Blayney, looking for a job. My brother lived here for five years and he only went to Scone three times. I'm the only man around here who is a local, who was working within a hundred-mile {about 160 kilometre) radius of here in 1950 and doesn't own his own place. Every man here now has come from someplace else.

LESTER TEAGUE

The speaker is Lester Teague, a 57-year-old boundary rider-cum-dingo trapper who tips the scales at 57 kilograms soaking wet. As his conversation indicates, he's been around the area all his life. We are talking of Glen Rock, a large sheep and cattle property situated north of Sydney between Gloucester and Scone. Glen Rock is slightly north of this line, in the Great Dividing Range. It is beautiful country, and rough country. Last year a 17-year-old youth died on the property while out pig shooting. He was unused to the country, he was driving and shouldn't have been, and his four-wheel-drive vehicle tumbled down a steep grade.

Lester lives in a cottage a few kilometres from the homestead. It is only discernible once it has been pointed out; Lester's wife doesn't have his enduring love for the high country and she only spends part of her time with him. She has a place in town where she can enjoy the amenities of civilisation.

What are Lester's living conditions?

I've got good facilities at my place. There's a Lister diesel generator for power, there's a proper toilet with water laid on for sanitary needs and there's a gas-lit water service. There's no phone, though. You earn close to $400 per week plus your meat. You just have to provide the rest of your food.

I never drink, never smoke and I don't knock myself around. I wear this kerchief all the time because it prevents sunburn, and I never take my shirt off.

I've been here on Glen Rock station four times. I first came in 1950, then in 1962, again in 1972 and now since 1985. Fourteen men staff this property. The hands last anywhere from one week to twelve months but the average man stays for three months.

We've got over a hundred horses; you start with a fresh one each day. This country is hard on a horse. You start out with two fresh dogs each day; you keep your own dogs here. Some men have six or eight; I have fourteen dogs.

I've known Ivan (Letchford) since 1950, when he worked for Grazcos; classing he was then. Some of these shearers have been with Ivan ten or more years. Considering that there's so much more wool, the shearers are doing a good job. They're not cutting the sheep or knocking them around. These sheep are so big because we've had thirty-odd years of super (superphosphate fertiliser) on the land. We used to average nine pounds (4 kilograms) of wool per sheep and now its up to 15 pounds (7 kilograms) per sheep.

The rich diet acts on sheep like gout on people, so they tell us. Before there was super we had no sheath rot. Now 4 or 5 per cent get sheath rot. (Sheath rot is a condition which particularly affects wethers. Lush green feed contains more nitrogen than the sheep can handle and it is excreted via the urinary system. 'Scalding' of the sheath results and opportunity for infection occurs. It is painful to the sheep, can be severely immobilising and even lead to death.)

There was no power here in the 1950s. It's been gradually put on from the 1960s. Living conditions are better now, there's no doubt about that.

This shearing shed was built about 1930 or 1932 from local timber which has been cut within a mile (about 1.5 kilometres) of the place. At first there was an old wooden press here. This is the third press. There was no electricity into the shed till I came back here in 1980. Before that it was generator-powered.

Blowflies are a problem here about seven months of the year—from September around till late April. Any given mob generally gets jetted up to two or three times per year. If you've got an inch (2.5 centimetres) of wool on your sheep the flies will start going into them. We probably drench them four or five times per year. The wetter it is the more they need drenching. It lasts six to eight weeks or up to three months if it's not too wet.

Anything from six to eight miles (about 10 to 13 kilometres) from the shed, we bring them in. Farther out than that and we set up a temporary race. A mob can be moved six miles in two hours but remember, no mob can be moved faster than the weakest sheep.

Back then the place was overrun with rabbits. There was less feed about, although there was more water in the creeks then. Early on they ripped many thousands of acres with a dozer but it wasn't really successful. Myxomatosis worked. It breaks out each year now.

Dingoes probably get up to a thousand sheep per year including the ones they eat and the others that they just chase and bite. If the dingo bites deeply the sheep usually dies because the wound gets poisoned.

Usually falling rain makes it easier to track dingoes for trapping. I don't take my own dogs along. I take my horse and put my traps where the dingoes are travelling. If I've got a bitch on heat I put some dirt from around her kennel with some dog droppings and I place the mixture about a foot (about 30 centimetres) from the trap.

I've been here two-and-a half years this time. The neighbour chap and I have got forty dingoes

WAITING TO BE SHORN.

between us during this time. Since they have four or five pups per time and half are probably bitches, they would have had two litters in this time. So we've prevented at least 160 young ones·from being born. We're winning slowly but certainly.

They're baited once a year too, out in the back country, so we've probably killed another twenty to fifty that way. At first I used to see two or three dingoes every week. Now I don't actually see a dingo but every two or three months.

5

As ONE *Man*

For three-quarters of a century, the shearers' union was the largest one in the country. A brief look at unionism in the shearing shed provides a window on Australia's political development and its pattern of industrial relations. The contrariness, the friction, the belligerence, the general strife that marks Australian labour relations have not just happened randomly. They are characteristics of a system that put prisoner against gaoler and shearer against squatter. We could expect little else under the circumstances.

The conditions for shearers in the last century could be best described as appalling. Work was conducted under provisos set entirely by the pastoralist; the shearer accepted what were often tyrannical rules or he didn't work. When penalties were applied, in all instances they pertained to faults of the worker, never to errors by the landowners. The rise of unionism in such circumstances was as inevitable as the sun's rising in the east.

Rules varied from property to property. While some employers were fair, compassionate men, others were harsh, rigid despots. Shearers worked according to contractual agreements and these could be quite oppressive. One from Victoria, in 1870, declares that bringing alcohol onto the premises and 'unfitting himself' for work not only causes dismissal but the shearer shall 'forfeit any and all monies which may be due him', regardless of whether the previous work had been done satisfactorily. Cursing, swearing and even loud talking were reasons for dismissal. All shearers were to work from sunup until sundown, except on Saturday when they would finish 'a little earlier'. Another agreement directed that work was to continue from 6 a.m. until thirty minutes before sunset. Even the most liberal of accords set the period at nine hours per day, the smoko breaks being granted against the shearers' own time. A South Australian agreement said that 'singing, whistling, swearing and noise' were prohibited. Another allowed the pastoralist to fine or fire a shearer using 'bad or improper language'. Yet another agreement fined a shearer the wages for shearing twenty sheep if he 'swore or used off-colour language'. The same contract called for the shearer to carry the sheep from the holding pen to his stand on the board. The penalty for not carrying the sheep was that the shearer was not paid for the sheep not carried. Considering the increased size of today's Merinos, that clause would be unachievable, especially when a hundred or more animals are shorn daily.

Sometimes meals were included in the contract, other times the shearer had to pay for his own. In the latter instance he purchased his food supplies from the station store at inflated prices. (There weren't any late-night convenience shops in the colonial bush.) Since the landowner bought the supplies at wholesale prices, selling to members of the shearing team at above-retail prices was a tidy way to increase profits. Moreover, the workers' meals—even if provided as part of the employment terms—were of basic fare.

When they did want something in the line of fresh fruit and vegetables they had to purchase these goods, commonly from hawkers who travelled the byways.

Housing conditions were terrible at best. Shearers' huts were primitive structures, unlined, often with no more than an earthen floor. Sleeping took place in tiered bunks which were supplied with grass- or straw-filled mattresses. Sometimes the huts were built without windows. Wicker bottles fuelled by kerosene comprised the internal lighting. What a foul-smelling, stifling dwelling such a shanty would have made through the long, hot summer. As a final affront, the shearer commonly had to pay the landowner for grass eaten by his horse while he was shearing the station's sheep.

One of the leading causes of dissidence was brought about by an innocuous entity called a 'raddle'. Being nothing more or less than a coloured marker, it was used on sheep by the boss to indicate those he felt were poorly shorn. Sheep thus raddled were deducted from a shearer's tally and not paid for. Even worse, in certain cases, the contract called for the employee to forgo payment for an entire pen if one sheep had been raddled. There is no denying that careless, negligent shearing took place. Indifferent tradespeople have always existed, today's society being no exception. And the employer had to have means to ensure that his flock was shorn properly. Nevertheless, inequitable and arbitrary application of the raddle was a fraudulent method of increasing the station owner's profit margin, at the same time foisting a degree of subjugation upon the shearer.

If disputes arose between landowner and shearer, and the disagreement was serious enough to reach the stage of adjudication by a magistrate, the usual authority figure was a squatter or other member of the colonial gentry. So we had a situation where shearers, who normally worked in what may be called difficult circumstances, likewise had to endure wretched living conditions and, as a final put-down, were bound by contracts which were imposed by the power structure of the times.

That's one side of the coin and, like all coins, there are two sides. Shearers were always comparatively well-paid workers. They could make £3 or £4 per week in the mid-1880s while other categories of shearing team members made considerably less. Stationhands worked in the same austere conditions as shearers and they received a fraction of the wages. A competent shearer could earn in one week what a shepherd struggled to make in two months. In many instances, shearers had their own cook, separate from the one supplied to the rest of the shearing team. Sometimes they were provided with segregated mess quarters. There is no doubt that shearers were a breed of outback elite, considered to be specialists of sorts. During times of prosperity their superior wage structure was easy to support. Again, during the gold rush when so many rural workers left to seek their fortunes on the goldfields, shearers who stuck with their jobs were generally well rewarded. Through the general time frame of 1840-1900, £2 per hundred sheep shorn was paid. (Oh, for such an inflation rate today.) During the 'bust' portions of rural Australia's perennial boom-and-bust cycles, a lot of property owners were squeezed until they were nearly strangled. Colonial records indicate that many did perish financially for there were economic, political and cultural forces at work beyond the simple matter of whether sheep would thrive on a certain property.

When hard times struck the man on the land, driving down his income, he wanted the shearer to share in the Sisyphean conditions by reducing his wages for the duration. This the shearers refused to do. Relations between shearer and squatter had been generally agreeable until the 1860s. Following a crash late in that decade, the association began to sour. Capital from the British motherland had to be serviced and maintaining overseas dividends required financial stringency in the Antipodes. The British version of class struggle appeared in Australia as the now familiar worker-management incompatibility surfaced. Comments Heather Ronald: 'The old time squatters and selectors had worked peacefully alongside men of all colours and creeds, and found the demands of this new breed of worker impossible to understand.' The men Geoffrey Blainey calls 'the princes of labour'—the shearers—wanted to hold onto their prosperity and keep their status among rural workers. And when one observes the lifestyles of the rich and powerful station owners, the shearers' demands seemed entirely justified. The problem was that not all the graziers were wealthy patricians. A good number were struggling to make ends meet in extremely difficult circumstances and this group could not afford to hire the princes of labour at regal wage rates. Antagonism and skirmishes were inevitable. So was the rise of unionism.

It is well known that during Australia's founding years, cheap labour was used extensively. There's not much cheaper labour than the free variety supplied by convicts. Even after they were freed, many pastoralists begrudged giving these ticket-of-leave men anything beyond their keep. They were hard men for hard times. But this situation could not last forever.

A shearer wrote to the Toowoomba newspaper in 1874, seeking colleagues to begin organising for a union. There was little success until William Spence became involved. Spence was a Scotsman from the Orkney Islands, arriving on Australia's shores at the tender age of six. He initiated his working career at the age of thirteen, beginning as a miner; later jobs included stints as a butcher and a shearer. Spence led the miners in 1878 when the employers decided to reduce wages; he called a strike, defeated the bosses and became a powerful force in the nascent union movement. He turned his attention to the wool industry and, in 1886, became the founding president of the Amalgamated Shearers' Union. The writing of William Spence, via Tim Hewat, explains the ready interest engendered in workers by unions:

> Unionism came to the Australian bushman as a religion. It came bringing salvation from years of tyranny. It has in it that feeling of mateship which he understood already. Unionism extended the idea, so a man's character was gauged by whether he stood true to union rules or scabbed it on his fellows.

Shearers were different from other toilers in the bush. They worked together, lived together, and took what little leisure time they had together. This propinquity enabled them to better organise themselves and therefore to better stand up to the group who they believed could well manage to pay them better, the landowners.

At first the scattered, ineffective unionists were powerless to accomplish anything of substance. Gradually, amidst setbacks and blunders, they gained strength. Disputes arose which, from our perspective a century later, sometimes appear sad or tragic, while at other times the major outcome was humour.

The first notable strike took place in 1886. Sir Samuel McCaughey was one of the biggest landowners in the country. He had properties in Queensland and New South Wales which together shore a million sheep. Four of these stations on the Darling and Paroo Rivers included Dunlop, Toorale, Fort Bourke and Nocoleche. Today, we tend to forget the tremendous scale of endeavour which took place in colonial times. To gauge the size of McCaughey's properties, reflect upon the fact that these four properties occupied 450 kilometres of Darling River frontage. The near-sovereign McCaughey and a number of other New South Wales graziers decided that the shearers had been receiving too much pay, so in 1886 they decreed that the rate per hundred sheep shorn was to be dropped from £1 to 17s. 6d. As would be expected, the men went on strike at the forty-stand Dunlop shed, one of the shearing season's early starters. Even though the station manager halted their supplies, the men, with support from colleagues, held out for nine weeks. It was long enough. The mustered sheep had to be shorn and so McCaughey 'paid the pound'. By the end of the year, 9000 men belonged to the union in Australasia.

Disputes increased and in 1888 there was a near riot at Brookong station. While watched by local police, union shearers removed non-union workers from the huts. The owner overreacted, sent for forty revolvers and ammunition, telegraphing that they were 'in a state of siege. Situation intolerable ... police authorities powerless.' When it came to court the magistrate agreed, even though not a single person had been harmed in any way. If one considers today's puny sentences for violent crimes, the penalties for what was nothing more than a labour dispute was astonishing. Three union organisers were jailed for two years and two others were sent down for three years.

Perhaps the charges were simply part of the tenor of the times. Again in 1888, two men were removed from a strikers' camp at Penshurst, Victoria, and accused of 'conspiring to raise wages, forcible entry, deporting shearers and assault and riot'. All the other charges fell by the wayside but 'conspiring to raise wages' earned the gentlemen a sentence of one year at hard labour. Quite astounding. If such a charge was to be effected today, the prime minister and most of the government would be locked up before the rest of us were gaoled.

1. GOOD STANDARD OF ACCOMMODATION FOR A SHEARING TEAM. AT FAR LEFT ARE TWO TOILET-SHOWER BLOCKS. CENTRALLY, THE OVERSEERS QUARTERS ARE IN FRONT OF THE SHEARERS' HUT WITH THE MESS AT THE RIGHT OF THE LONG BUILDING. THE SMALL BUILDING AT FAR RIGHT IS THE MEAT HOUSE. 2. MANY SHEARERS HUTS DO NOT HAVE VERANDAHS, WHICH ARE A REAL BONUS IN HOT WEATHER. 3. TYPICAL SHEARER'S ROOM WHICH MUST BY LAW MEASURE A MINIMUM OF 100 SQUARE FEET. EACH YEAR A 'HUT INSPECTOR' FROM THE DEPARTMENT OF LABOUR AND INDUSTRY (NSW) CHECKS THAT ACCOMMODATION IS UP TO STANDARD. WHERE THERE ARE MORE THAN EIGHTEEN ON THE SHEARING TEAM A 'HUT-KEEPER' MUST BE EMPLOYED. THIS PERSON SWEEPS OUT THE ROOMS, CLEANS THE TOILETS, KEEPS HOT WATER GOING AND MAY EVEN BE COAXED BY THE COOK TO PEEL A FEW SPUDS.

Such actions had predictable consequences. By late 1889, in New South Wales, Victoria and South Australia, there were 2400 union sheds and in the following year 20 000 unionised shearers.

William Guthrie Spence taught Sunday School in a Presbyterian church as well as occasionally preaching from the pulpits of the Primitive Methodists and Bible Christians. He was a member of the town council and considered to be a fine family man. Nevertheless, the avowed Christian and his Amalgamated Shearers' Union had some strange ideas about the 'brotherhood of man'. Yes, it was a period in history when a White Australia Policy had many adherents. Yes, there was keen competition for jobs. And finally, yes, the men were afraid that the Chinese in particular would work for lower wages than the local Australians. So, when the union held its annual meeting in 1890, they decided to withhold membership from Chinese and kanakas, the latter being South Sea Islanders who were sometimes press-ganged into employment in tropical Australia.

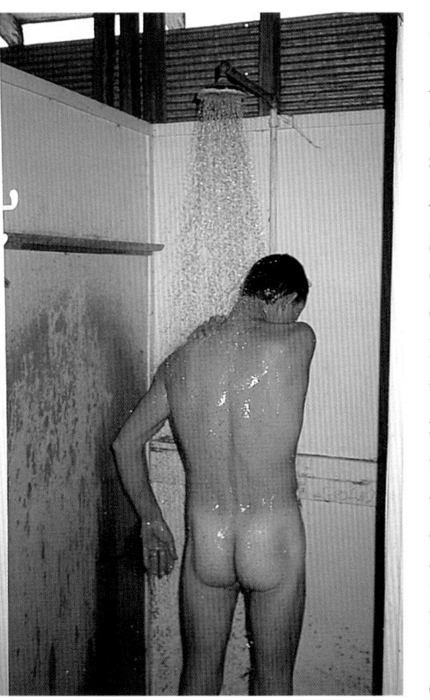

4. SHEARING IS HARD, DIRTY WORK BUT SHEARERS ARE VERY CLEAN PEOPLE. WHILE SHOWER FACILITIES ARE QUITE BASIC, THEY ARE WELL USED.

Patsy Adam-Smith says it was by-law 62 which excluded the Asians and kanakas from serving as either shearers or shearers' cooks. It seems Maoris were all right but Aboriginal workers weren't exactly given warm welcomes. Graziers didn't go along with this white employment policy. Why not? Because their faith was true and they believed in the brotherhood of man? No! Because they felt it was their right to hire and fire whoever they chose. John Merritt explains that the 'Agreement of 1891' between unionists and pastoralists found the latter group vowing to 'use its influence' to prevent Chinese and kanaka employment.

It's an ironic little twist of national history that while the downtrodden suppliers of labour were seeking justice, they were themselves scorning another group who sought fair play.

A major strike began in Queensland in January

NOT THE PRIME MINISTER'S INNER MINISTRY BUT A SHEARING TEAM ON A 'SMOKO' BREAK.

1891 and before long other colonies and unions became involved. There were many violent clashes in the streets of Sydney, Melbourne and Brisbane between troops, police, strikers and their supporters. Wharf labourers in Sydney and Melbourne attempted to prevent non-unionist shorn wool from being loaded on ships bound for Britain. As Geoffrey Blainey put it, 'Here was a trial of strength between unions who were sensibly trying, where possible, to make membership of a union compulsory and employers who were sensibly trying to control their working costs and their future.' The wool arrived in Sydney during August, 'scab' labour was plentiful and the strike ended, a total failure, by early November.

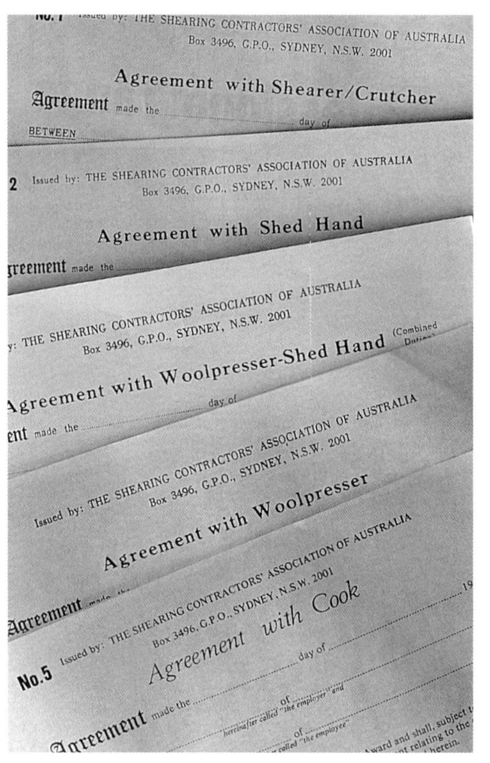

A SAMPLE OF A SHEARING AGREEMENT AS WELL AS THOSE FOR OTHER TEAM MEMBERS. THE SHEARER AGREES TO SHEAR A MAXIMUM NUMBER OF SHEEP PER SHED WHILE THE CONTRACTOR AGREES TO PROVIDE A MINIMUM NUMBER OF SHEEP FOR SHEARING. AGREEMENTS PROVIDE EMPLOYEE AND EMPLOYER WITH A GUIDE TO THE EXTENT OF THE JOB AHEAD.

There were serious shearers' strikes in both 1891 and 1894. In neither case was any substantial gain made by the union movement. Considerable open, general strife occurred during the final decade of the nineteenth century but a severe drought during the same period once again meant hard times for shearer and squatter. Troops were called out in Queensland, for it was the state closest to anarchy. Fortunately, talk of civil war died down as everyone attempted to get by.

Eight or more shearing sheds were burned down in Queensland. At one station, seventy-five mounted men set fire to hectares of pasturage; good luck and good station hands brought the fire under control. Although it sounds far-fetched, it was reported that men went so far as to capture kangaroos, attach 'burning materials to their tails' and release them to fire the countryside.

In *The Great Days of Wool*, Joan Palmer tells a couple of anecdotes about efforts to protect a shearing shed from arsonists by encircling the area with bells hung on wire:

> Hearing bells tinkling like mad in one spot on a dark night, a trigger-happy jackeroo fired in that direction ... next day they found he had shot dead a 500-guinea ram that had been put in a small paddock near the shed. At the same shed, one of the men on duty was doing his rounds when he saw a sinister shape moving. When he challenged it and there was no reply he shot it, only to find he had drilled a hole through his own overcoat which he had hung up when commencing his night's work.

One ill-fated story of misfortune ends with a splendid line. Tim Hewat writes of a fire destroying Boonoke homestead in 1888. 'One eve a union organiser arrived and Falkiner ordered him off the place.' An argument took place while each man was mounted and Falkiner charged the man, driving him into a billabong. After dinner that same night, the house was burned down. There were horse tracks right up to the point where the fire started. Naturally enough, the union organiser was the prime suspect. Hewat concludes: 'But there was no real evidence against him and he became a politician.'

In 1894, the Amalgamated Shearers' Union officially changed its name to the Australian Workers' Union (AWU) becoming, as mentioned earlier, the nation's largest union. The AWU retained this position for nearly seventy-five years. It was a time of reduction from the huge property days to those of rationalised stations, from great flocks to the more modest sizes prevailing to this day. In the Western Division of New South Wales, sheep numbered over 13 million in 1894 and by 1900 there were less than 5.5 million remaining. A typical property in the Tibooburra area which had shorn 130 000 in 1899, had shorn 101 000 in 1900 and that figure was reduced to less than 26 500 by 1902. Drought, erosion, rabbits and sheep diseases permanently changed the wool industry. By 1907, employer and employee had had enough of internecine warfare. A shearing pact emerged, as did the first shearing award. Peace spread across the paddocks of Australia.

In reviewing the failed strikes of the 1890s there was an unexpected aftermath which went well beyond the consequences of garden variety industrial unrest. Author Richard Magoffin best summarises the result of the shearers' union defeats:

> Observing that the employers had all the governmental agencies, including the army, on their side, the labour leaders determined to gain control of the government by peaceful means. Accordingly, the labour parties were formed, and the unions entered politics directly. In the twentieth century, they have played a decisive part in Australian affairs, having been elected to office more often than the conservative parties. All this seems to have stemmed mainly from the failure of the strikes. The year 1895 may be regarded as a turning point in Australian history.

By and large, the job got done up until 1982. It was then that the infamous 'wide comb dispute' brought fear and loathing to the shearing shed. Once machinery replaced the blades a narrow comb had been used for decades. In fact, as far back as 1926 the Federal Pastoral Industry Award had decreed in Clause 32 that the comb could not be wider than 63.5 millimetres (close enough to 2-½ inches). The shearers had not sought this limitation; it had been requested by the graziers to lessen injuries to shorn sheep, even though there was not another country in the world with such a rule. Then, in the 1960s, a wider comb was developed by the major manufacturers. The widest were 86 millimetres (close to 3-¼ inches) across, possessing thirteen teeth instead of the usual ten. The wide combs gained support almost immediately across the Tasman in New Zealand. There, Merinos form a small portion of the total sheep population while the majority sport a sparser fleece which local shearers sometimes refer to as 'fairy floss'. Kiwi shearers, including a number of Maoris, took the wide combs to work in the vast

Flash Jack from Gundagai

I've shore at Burrabogie, and I've shore at Toganmain,
I've shore at Big Willandra and on the old Coleraine,
But before the shearin' was over I've wished myself back again
Shearin' for old Tom Patterson, on the One Tree Plain.

Chorus:
All among the wool, boys, all among the wool,
Keep your blades full boys, keep your blades full.
I can do a respectable tally myself whenever I like to try,
And they know me round the backblocks as Flash Jack from Gundagai.

I've shore at big Willandra and I've shore at Tiberoo,
And once I drew my blades, my boys, upon the famed Barcoo,
At Cowan Downs and Trida, as far as Moulamein,
But I always was glad to get back again to the One Tree Plain.

I've pinked 'em with the Wolseleys and I've rushed with B-bows, too,
And shaved 'em in the grease, my boys, with the grass seed showing through.
But I never slummed my pen, my lads, whate'er it might contain,
While shearin' for old Tom Patterson, on the One Tree Plain.

I've been whalin' up the Lachlan, and I've dossed on Cooper's Creek,
And once I rung Cudjingie shed, and blued it in a week.
But when Gabriel blows his trumpet, lads, I'll catch the morning train,
And I'll push for old Tom Patterson's on the One Tree Plain.

DAY WEDNESDAY DATE 7 JUNE '89

Stand	SHEARER	COUNTS				OUT	Day's Total	Previous Total	Total to Date
1	F. Bowen	48	45	47	40		180	1502	1682
2	P. Ryan	39	40	43	38		160	1476	1636
3	D. Murray	32	35	38	38		143	1139	1282
4	J. Bridge	15	14	23	28		80	723	803
5	H. Phillips	58	42	55	45		200	1568	1768
6	R. Lang	62	48	48	49		207	1581	1788
7	J. Murray	38	45	47	37		167	1490	1657
8	P. White	28	32	38	23		121	900	1021
9	N. Thompson	27	35	35	41		138	1137	1275
10	H. Hammett	23	25	28	27		103	921	921
11							1,499	12,334	13,833
12									
	TOTALS								

A CLOSE-UP OF THE TALLYBOOK, REGULAR READING MATERIAL FOR EVERY SHED spaces of Western Australia. A thickening pot was further congealed by New Zealand mail-order companies who decided to market Australian-manufactured wide combs to Australian shearers. More ill feelings arose through the animosity of the Australian Workers' Union hierarchy towards cross-Tasman shearers, especially those who were Maoris.

Heated debate took place in Arbitration Commission hearings throughout 1982. The AWU representatives stated that the wide combs were more likely to injure sheep; they would be harder to push through the fleeces, thus harder on back, wrists and arms; that there would be increased cuts on and about the hands; finally, there was vague, non-specific opposition on the basis of industrial health grounds.

Our own Ivan Letchford represented the Shearing Contractors' Association of Australia before Commissioner McKenzie. The following is a sample of Ivan's testimony:

> We do not accept that the extra three teeth in the comb is the real issue involved, but the fear and we believe a very real fear of the way the award conditions are being eroded by the imported shearers and contractors using the wide comb. Why the union does not make that their prime target, we cannot understand. We hear of breaches of the award constantly.

Again, a short while later:

> Thanks to the union efforts, we now have good conditions, reasonably good money, good machinery and each change has been for the better. Shearers have used 64 millimetre combs or less during the last fifty years, shorn big tallies and did the job well. But how do we know the wide comb will not be a further improvement until it is freely tried? Many reliable shearers are using the wide comb and they say they get the same number of sheep easier.

Commissioner McKenzie agreed with these sentiments, as did the full bench of the Arbitration Commission, in March 1983. The wide comb was given legal approval and the union immediately went

SHEARER PEGGING OUT WASHING ON CLOTHESLINE. TO AVOID SKIN INFECTIONS IT IS IMPERATIVE FOR SHEARERS TO KEEP THEIR WORKING CLOTHES CLEAN.

CULLING FLOCK EWES. THIS MEANS THAT THE LESS PRODUCTIVE WOOL CUTTERS ARE REMOVED FROM THE MAIN FLOCK.

on strike in protest. Substantial personal injury resulted from gung-ho emotions, needless property damage transpired and seething tempers kept the countryside at boiling point. Nevertheless, within a few months everything settled down and today's estimate of the use of wide combs approaches 99.9 per cent.

A shearing contractor told Tim Hewat, 'You would have to cut their arms off to get the wide comb out of their hands now.' Moreover, the union has to a degree left the shearers to their own devices.

A Cook Discussing the Union

The men are so weak. You'd get word that the union organiser was coming around. The men would say, 'When he gets here we'll tell him this and we'll tell him that'. The minute the union organiser set foot on the board they'd go to water, the lot of them. Whatever the union organiser said, they'd go along with it. They'd fork over their money when they didn't really want to and that would be the end of it. They have only themselves to blame if the union is weak or poor.

That's how the men are: weak, weak, weak. The union never did anything for the kitchens. You never got improved stoves, for example; too often they were still the wood ones.

I worked without a union ticket for a year and eventually they got me. I said to them, 'The only reason I'm going to be a member of this union is so I can tell you lot of bastards what I think of you. This is my ticket for the freedom to voice my opinion.' After that they never came back down to the kitchen to see me. Once they got my money there wasn't any point in them coming down to see if I needed anything or if I was happy with my status.

The Union

The following is a collection of quotes from a number of shearers discussing their union.

The unions are not doing the right thing. We just don't see an organiser now. You see one occasionally when the union ticket is due. Years ago the rep would come along, have a meal with you at night, you'd have a union meeting after the meal and the rep would sleep in his car. Today you don't see them at all. They start about 9.30 in the morning and they don't camp out any more. I haven't seen one since I bought my last union ticket.

The pastoral workers are the men who kicked off the AWU and now we're getting a kick in the pants. It used to be it was all shearers in the union and now they've even gone to shop workers. You never see them. The only time we see them is when they want $210 (annual dues).

Years ago, anyone coming around without a union ticket would be sent away. The union was so very strong. Today the situation is—I don't know how to put that big word—ludicrous. Anyway, it's just a heap of crap. Years ago it was entirely different. My father and Banger's father fought for this union, and fought very hard, I might add. Today they don't even look after their own people who really kicked the game off. It was kicked off as a pastoral union for shearers—not bloody store workers or miners—and now they have got so big they can't even look after us. Instead of looking after 10 000 workers, they're looking after about 50 000.

Years ago—I'm not that old either—before you started a shed you had breakfast and then you voted in a rep. The rep saw the cook and made sure she had all the necessary utensils, pots, pans and dishes. He went around and had a look at the wash house, made sure there was a hand basin there and tubs for doing your washing at night. Then you had a show of tickets and made sure everyone was financial. Only then would you go up to the shed; you might not start until the second run at ten o'clock. There were times when we didn't start for two days if some blokes didn't have their tickets or weren't financial. If the place had a bad name from the year before, we made sure all was well. If they sent out old, cracked crockery from the homestead that was no good to us, we'd tell them we wanted new plates and a knife and fork per man, or the cook needs this and that.

In those days you always voted a rep in. You might sit down and deliberate for three or four hours who was going to be your rep. Now, he's just always the number one stand.

A Shearers' Parable

Dennis Ryan is a third-generation shearer who has left the boards and runs a successful business supplying requisites to his former shearing colleagues.

This is a parable but it is a true parable. My grandfather came from Ireland. He settled in Glebe with his wife and a couple of children, which he added to greatly later. My grandmother started a boarding house while he went to find work. He strikes a bloke in the pub who says, 'Why don't you come to the bush with me and together we'll get work?' And so he did and they became shearers. Now my grandfather did that in 1890.

At that time my grandfather was the rep and he came up to the boss and said, 'What do you reckon about us?'

The boss replied, 'Oh, you are good, very good.'

'Well,' said my grandfather, 'We don't think much of you.'

'Why don't you?' asked the boss.

'Because we want a tap,' said my grandfather.

'You want a tap? Why can't you be like everyone else and go down there to the creek and have a wash?'

'We're not like everyone else,' said my grandfather. 'We're shearing your sheep, we're working very hard for you, we're doing a special job, it's greasy and dirty work and we need a tap. We want you to put some plumbing in so that we can actually go and do the work you want us to do and we can finish up like ourselves at the end of the day.'

The boss couldn't believe it, but everywhere they went they told the boss that they wanted a tap. So, in actual fact, the foundation of the Labor Party in the big strike of 1890 was all over a tap. They got their tap.

My father would catch the train to Moree. He'd buy a pushbike and he'd ride all over the back country until he got to the shed. Remember that this is the next generation.

He'd ask the boss, 'What do you think of us?'

The boss would reply, 'Oh, fine, you're very good.'

'Well, we don't think much of you,' my father would say.

'Why is that?' the boss would ask.

'Because we want a bed,' my father would say.

'A bed?' the boss would reply. 'Shearers want beds? You gotta be kidding. You get a bike at Moree and you ride out here for bloody three weeks, sleeping on the ground all the way and you get here and you've got the hide to ask for a bed.'

'Yes,' my father would say, 'I'm working for you now and this palliasse cover is filled with straw and that's not good enough. We want a bed.'

So there was a big strike over beds, until shearers got their beds. Now I go out shearing near the end of the war, oh it's about '43.

I say to the boss, 'What do you think of us?'

'Oh, you're bloody marvellous,' he says.

'Well, we don't think much of you. We want a refrigerator,' I reply.

'A refrigerator,' the boss laughs. 'If you get a refrigerator all you'll do is cool your bloody grog in it.'

'We have an understanding,' I say. 'There will be a rule in the shed that there's no grog to go in the refrigerator. It was my turn to kill the sheep last night, it was 100 degrees all night, so this morning the chops were bouncing all around the kitchen, rigour mortis hadn't even set in. That's why we need a fridge.'

The boss said, 'I've never heard of anything so ridiculous in my life.' So, we went on strike for six

SHEEP GRAZING ON LOVELY, ROLLING COUNTRYSIDE.

months and we got ourselves a fridge.

Under today's circumstances what we asked for is regarded as unbelievably normal. We got those things and they are what is termed 'part of Australian life'. They are things we went to a lot of trouble to get. You get nothing unless you ask for it and nine times out of ten it is still refused. So we got these things. Now the country is riddled with people who go out to the shearing sheds and they don't want a tap, they don't want a bed and they don't want a fridge. The Maoris go out with a motor car now and sleep in the car.

So we lost the whole bloody lot in eighteen months. What it took us a hundred years to win we've lost in eighteen months.

Oh, that's all right for 25-year-olds. When I was 25 years old I could do anything, and anything was all right for 25-year-olds. However, an industry built on 25-year-olds is an unhealthy industry. It's got to be built on 50-year-olds or there is no future for the industry. It's knowledge that creates an industry and if you are going to have only healthy, strong new chums, it's a bosses' industry.

Shearing doesn't have any romance any more. We used to love a good strike.

I tell people now that I've got a real sharp wether's horn at home on the mantelpiece along with a jar of maggoty wool. Every time I want to go shearing again I just get a few lungfulls of wool and rip myself up the ribs a couple of times with the sharp wether's horn and it cures me for at least a fortnight.

ABANDONED OUTBUILDINGS IN MARGINAL WESTERN REGION.

6

MODUS *Operandi*

Today's shearers do know about the golden days of sheep and sheep stations but in general they work on a much smaller scale. Before the turn of the century, there were several properties ranging in size from 200 000 to 800 000 hectares.

Sheep carried on a single property could number as many as 400 000 animals. There were a dozen flocks numbering over 100 000 sheep as late as 1920. Now the picture has changed considerably. Depending on the location of the property, average flock size varies from 2000 to 7000. Accompanying this change has been a gradual transformation in the arrangement of shearing. Traditional shearing of the Ivan Letchford variety, where an entire shearing team including one or more mess staff move onto a property, living there until the collective shearing tasks are completed, has declined. It has been replaced by a form of neighbourhood endeavour. This new method is frequently called suburban shearing and the men engaged in this form of work are sometimes labelled, for obvious reasons, cut-lunch shearers.

Shearers have worked piece-rate since the 1840s. Yes, they had largely swung over to union allegiance but they were still employed singly, one man at a time. An abrupt innovation occurred at the end of the last century. Departing from the old ways, the appearance of the shearing contractor brought about changes which were jointly beneficial to the interests of the shearer and to the pastoralist. Contractors provided a form of secure, regular work for the bladesmen. For the station owner they metaphorically removed the kinks from the wool-gathering task. As well, contractors assumed the role of middleman in relations between the provider and the hirer of labour. This simple interposition enabled the wool industry to operate on a much more workable level. Here we are, nearly a hundred years later, and the system of shearing contractors continues to work so well that there are no major alterations on the industry's horizon.

Early 1899 witnessed the placement of the first public notice by a shearing contractor. L. McLean and Company, who traded as the Victorian Shearing Company, advertised for men to work on a shearing team. Contractors became responsible for shearing approximately 90 per cent of eastern Australia's sheep in just twenty years. Patsy Adam-Smith comments on the function of the contractor, positioned between pastoralist and shearer:

> By the very nature of things, what woolgrowers want cannot be what shearers want, and vice versa. But the contractors attempt to be fair to both. They are in a position to see both sides of the matter and it is because of them that the industry flows so smoothly today. It isn't that alone that excites admiration, however; it is their total involvement. Perhaps they love the work

they are doing—there seems no other reason for their killing hours, hard, lonely travels on bad roads, on call twenty-four hours of the day.

While contractors may have been a boon to the then beleaguered wool industry in the early 1900s, another phenomenon changed shearing life drastically. The universal ownership of motor vehicles progressively altered shearers' lifestyles as much as it affected the general population. Where formerly the team was kept close because transport simply did not exist, now shearers frequently go home on a daily basis. Former shearer Dennis Ryan comments:

'Shearing sheds now have become local jobs. They think nothing of driving fifty miles or so (about 80 kilometres) to a shed in the morning. They hardly ever have their own dunghill. Why, they are that busy on the weekend doing jobs for the missus that they don't even have time to go to the pub for a few beers!'

Ivan expands on the effects of suburban shearing:

> There's been a great change in shearing over the past twenty-five to thirty years. We've lost a good deal of mateship in this time. Men used to come to the big sheds year after year after year, with a strong sense of unionism all the way. They did the right thing by one another. When they arrived on a job the union rep was put in, a committee was put in, and they made sure that the shedhands were represented. The shed was then run as a unit, perhaps as a small village would be run. Rosters were put up to do the toilets, more rosters for the showers, etc. It was all done properly. We had no motor cars so that when we arrived at the shed that's where we stayed until cut-out.
>
> At night we had fires, inside or outside. We sat around spinning tales and yarns, sometimes we had card games and sometimes we played two-up. On weekends we had organised cricket matches.
>
> Suburban shearing has been a dreadful thing for the industry. They get in their motor cars each night and drive home and there's no thought at all for the other bloke. There were men who were known over the length and breadth of Australia. Blokes were spoken of and discussed around the fires at night. Through suburban shearing we've lost many of our most colourful identities.

The following is a collection of comments from some shearers. These are not cut-lunch shearers, these are traditional shearers so we know in advance what method of work they prefer. Also, these are mature men, not those in the bloom of youth for whom no task is unduly tiring. This is how they feel about suburban shearing:

> You're up too early in the morning, you got to cook your own breakfast, pack your own lunch, then drive an hour or an hour and a half to work. You finish and drive home, have your shower, have a few beers, cook your own tea, wash up, wash your Esky, get to bed and it's half-past nine or ten o'clock. Then it's up again at 5 o'clock next morning. When you go to a camping shed you're there on the job, all meals are supplied, there's no worries. You can shear more sheep and get more rest.

> It's the hours you have to travel. Counting in your smokos you normally do ten hours in a day but with suburban shearing it's more like fifteen hours in a day. Besides, you're doing a cook out of a job. Suburban shearing takes the traditions away.

> You can shear more when you are camped out. When you are older you can do more sheep when you go to bed at 8 o'clock.

> When I was younger I tried suburban shearing for a few years. I was driving a hundred miles (about 160 kilometres) there and a hundred miles back. It didn't matter because I didn't give a bugger then. I could get suburban shearing now, but I don't want it.

There is another way to classify shearing and that is in relation to the economics of the undertaking. The pastoralist hires a team to do the job on the basis of either a contract price or a cost-plus structure. In contract shearing a price is quoted to shear the sheep that have been delivered to the shed door. Included in the price is the handling of the wool, its classing, pressing into bales and being weighed and labelled. A quoted fee is a firm price. If it rains that's tough luck, the contractor stands by his price. If there are equipment breakdowns, that's more tough luck. It doesn't matter if the contractor has to wait out two weeks of rain while on a property, that's his loss. On the other hand, sunny skies during an assignment means a job done without delay and a brighter appearance to the profit column in the financial records.

Cost-plus shearing involves the same services provided to the owner but there is no fixed price guarantee. In this instance the contractor furnishes the essential product labour—and he receives, say, 20 cents per head above his costs for providing the services. Ivan comments, 'Cost-plus employers can put on "extra" staff if they wish and they are usually not as lean an organisation.' Cost-plus is quite commonly seen in suburban shearing.

As far as overall cost is concerned, Ivan says, 'My price could be more to start with but it's often less at the bell. A lot depends on the weather.' Today many contractors operate on a cost-plus basis. Why? Ivan says that prior to the current award they didn't have the resources to stand a good 'knock' from a dose of bad weather.

One would expect that shearers would be paid in cash since they are so often situated long distances from town. As it turns out, cash is never a form of payment for a shearing team, cheques being the normal method of settlement. Contractors' cheques are well accepted around the countryside, even by publicans and storekeepers. Ivan recalls days gone by: 'We'd go to the pub, even if it was the middle of the night. The publican would open up to cash the cheques. If he didn't have enough money I'd go to find the bank manager. I remember doing this many times. The manager would get the accountant to open the bank, even if it is was 10 o'clock or 12 o'clock at night. Once the men had their money, often they'd go straight back to the pub. No, there's not any cash in the shearing industry.'

How much are the men paid to shear a sheep? The question is almost in the same category as 'How long is a piece of string?' They receive different amounts for various sheep. There is an award, the federal Pastoral Industry Award, which governs the sum paid per hundred sheep shorn. Included in the award are flock sheep (wethers, ewes and lambs), rams, ram stags, stud ewes and their lambs, and double-fleeced sheep of the flock, ram and stud ewe varieties. The award rates vary from approximately $150 to over $300 per hundred. That is the rate for 'not found' shearing, which means that no meals are included in the working arrangement. A 'found' worker is one whose employer provides his food; in normal circumstances a certain amount of money is then deducted from the labourer's weekly pay packet. There's also a percentage payment to be added should an employer require any hand shearing to be done and another if a shearer is required to provide his own stud combs. A sample reply when a shearer was asked his rate: 'It's $151 for a hundred flock sheep, $189 for stud ewes and $302 for a hundred rams.' By way of example, rounding off the rate to $150 per hundred for shearing wethers, a novice shearing sixty sheep a day earns $450 in a five-day week. Similarly, the steady hundred a day shearer earns $750 per week, a good fast shearer doing 150 per day earns $1125 per week, and the uncommonly talented individual shearing 200 a day is rewarded with $1500 per week. Decent money? Yes. A noteworthy wage? Far from it, especially when one considers all the extras (pension plan, holidays, long-service leave, etc.) present in so many government and private sector jobs.

The shearer is a bastion of free enterprise in a semi-socialised economy. He may be heavily unionised but this fact is largely irrelevant to earning power in the wool industry. If he doesn't work, he

SHORN SKIRTINGS BEING READIED FOR BINS, PRIOR TO THE PRESSER PUTTING-UP THE WOOL INTO BALES.

doesn't earn. Moreover, the shearer does what can only be described as exhausting, back-breaking labour all day long. Percentage-wise, there's no longer much of the workforce that can lay claim to Herculean effort as part of their daily routine. The single factor which is absolutely amazing is how shearers can work so much of the time doubled over, literally bent in half. Hunter pastoralist Ron Taylor salutes shearers.

> If you are prepared to work, I don't think there is any profession today that is better than shearing. If you wish to work you get paid for working and if you don't wish to work you don't get paid. If this applied to all industries, Australia would be a lot better off. Shearers are paid on productivity whereas, with most union jobs, if you are over twenty-one your rates are the same. It doesn't make sense. The bad worker gets paid as much as the good worker.

Shearers 'sign on' before beginning an assignment at a shearing shed. This sign-on process actually documents the fact that they are beginning a new job. The contractual agreement is signed by both employer and employee and witnessed by a third party. Both sides agree to abide by certain commitments during the time of the shearing engagement. The contract guarantees a minimum and maximum number of sheep to shear. The maximum cannot exceed the minimum by more than 25 per cent. If the contractor is under in count, he has to make a payment to the shearers. As for the shearer, he cannot 'pull-out'—that is, leave before the job is done. Shedhands also sign a binding agreement. They are paid so much per 'run', not found. An experienced adult earns $29 per run while a youthful junior earns $20 per run. Thus, an experienced adult shedhand working four runs per day for five days a week would earn $580 per week. The employer would deduct approximately $126 per

FLEECES ARE SO TIGHTLY PACKED THAT A BALE MAY END UP WEIGHING 200 KILOGRAMS.

week if this shedhand was found.

For years the cook was hired on a weekly basis but this has now been altered to a daily basis with a wage guarantee of $128 per day. The cook's income depends on the size of the shearing team because he/she is paid $9.86 per day per worker, or $69 per seven-day week.

As for the cost in running a mess, not found workers divide the expenditure for the food amongst themselves. For example, ten men work ten days; the food costs $400 for a hundred days of work. This cost is divided by the total days worked and translates into a figure of $4 per man per day for rations. The team's meat, in the form of freshly-butchered sheep, is provided by the pastoralist as a part of the working agreement.

A discussion of money and shearers leads to the topic of indebtedness. Shearers commonly borrow money from their contractor-employer. The thinking behind the debt is that it assures the individual of a job at the next shed. Ivan says, '40 per cent of the men owe you money at one time or another'. He gives a graphic account illustrating the psychology of financial obligation. 'In the pub this day, a man bit me for a pound. Before I could answer, Angus turned round and smacked this bloke with a solid backhander, square in the mouth. As the blood ran out of his lip, Angus said to the chap, "That's the man I work for. Now if he lends you a pound he's gotta give you the job to get his pound back ... and I could lose mine." That was his form of reasoning.'

Another major change in shearing life is the way in which team members find work. Today's medium of communication is the telephone. A group of shearers explain the methodology of staying employed:

> We just know Ivan, when the sheds are going we give him a call. That's what the telephone is for. Or, if a contractor is pushed, he may give you a call.
>
> You take August or September, you wouldn't get a spare shearer floating around here anywhere. Once spring starts you can pick and choose your work.
>
> Ivan's got work all through the winter. He's one of the select few that has work when there's nothing going on. He'll be getting heaps of phone calls now. After July he will have to do the phone calling. They're chasing him now, looking for work. It will be different in another three months.
>
> I was with Ivan for eight and a half years before and we were never without work. We did lose a few sheds though, when some of the cockies went over to wheat.
>
> If you are lucky enough to have the money to get the phone on, you put your name in the book as 'shearer'. A cockie won't drive 50 miles (about 80 kilometres) to look for you but he will look in the phone book. This is the idea of the phone. It works the other way, too. You look in the book and say to yourself, 'There's old Gordon, maybe he needs men,' and so you jump on the phone. You might ring Quirindi or Collarenebri. You always look for someone who has got continuity

of work. I virtually work for just the two contractors now.

It wasn't always a matter of telephonics. Ivan paints an attractive word picture of how shearers—and jobs—were found in the earlier days of his career.

Years ago you used to be able to get on any train—but especially the North-West Mail—to Moree or Walgett, and when the first stop for refreshments came up (at Gosford) two or three shearers would get on. More would get on at Newcastle. It was never any trouble picking one up if you were short for a job. There used to be dozens of shearers living in Sydney. Carrying their swag they also cleared land, made hay, did stump-grubbing, seasonal work, harvesting and the like. The Hotel Sydney was a great meeting place for shearers. You could go there and find a shearer any time of the day. The Hotel Morris in Pitt Street was in the same category. Shearers had their favourite hotels in the southern end of Sydney, while in the northern, waterfront end, the farmers and graziers had their favourite haunts. They'd go to the Hotel Metropole, Aaron's Hotel, Hotel Australia and Usher's Hotel. Year after year, they would congregate in their favourite hotel, often staying in the same room each time. I won a lot of contracts at those old hotels. For a while I nearly could have run my business from them.

The shearing 'season' was originally just four months a year. It expanded to six months and now lasts all year round except for the Christmas holiday break. Traditionally, the season began in Queensland prior to the other states. Dennis Ryan explains how shearers got into the game during the 1940s:

We used to go to Cunnamulla on the first Monday in January for the start of the shearing season. You needed no money and you had no expense in getting started. If you were here in Sydney and you were broke, you could come to our hotel and find out where they were shearing. Someone would be leaving for Queensland and you'd get a ride with him. More than likely, if that fellow knew you or knew someone who knew you,

BRANDING THE BALES IS LITTLE DIFFERENT NOW THAN IT WAS A HUNDRED YEARS AGO.

he'd speak to the contractor for you and get you a pen. When you got to the job you said to the contractor, 'I've got no money, will you advance me half-a-dozen combs and cutters?' They wouldn't last very long, but they'd get you started.

Now Cunnamulla's not the beehive of work it used to be. We would go to Cunnamulla and fill up all the hotels. In those days everybody came to the pub. There were twenty of us living on the veranda. If you drank at the hotel you weren't even charged board. That was what it was all about.

Now, if someone said we can get you a job in Charleville, for example, where do you find out about it? There are no contacts in Charleville, there's no one leaving for Charleville tomorrow, there's not a surge of people with a convoy of cars heading for Charleville.

Then, year after year, I'd go to Mascot Airport the first Sunday in January. Butler Air Service used to fly Sydney-Bourke-Cunnumulla-Charleville. We'd leave here at 6.30 in the morning and be in Charleville for lunch. We'd have lunch at one of the hotels, then six of us, sometimes more, would ger a taxi—it depended on how many the taxi would hold. We'd drive out 280 or maybe 300 miles (about 480 kilometres) to the shed on the Sunday afternoon and you'd start work at 7.30 on Monday morning. It would be 115° (about 46°C) in the shade and bloody

ABOVE LEFT: FOUR CONSECUTIVE BALES FROM GLENROCK'S CLIP. BESIDES THE PROPERTY IDENTIFICATION, BALE INFORMATION INCLUDES THE CLASSER'S NUMBER, CLASSIFICATION OF WOOL AND BALE NUMBER.

ABOVE RIGHT: PRESSER, KEVIN WEATHERALL, WITH HIS EFFORT FOR A DAY. AS A GENERAL RULE, THE PRESSER CAN PUT UP ONE BALE PER SHEARER PER RUN PLUS THE ACCUMULATION OF THE BELLIES AND LOCKS OF ALL THE SHEARERS IN THE TEAM. THUS, IF THERE WERE SIX SHEARERS THERE WOULD BE SIX BALES PER RUN, FOUR RUNS PER DAY, THE TOTAL BEING TWENTY-FOUR BALES PLUS THE BELLIES AND LOCKS.

kookaburras would be dropping out of gum trees, gasping for breath, and we would be shearing a mob of sheep for the day. You were there for the first four months of the season and then you shifted.

Ivan explains why shearing changed from several months in duration to all-year round employment. 'New South Wales was zoned in those days. We went to the western district in late May and June for the winter shearing and we went to the north-east and Monaro regions for the September-December spring shearing season. What made it a year-round event? The advent of suburban shearing was a major factor. So was the cutting up of big holdings and the arrival of the soldier-settlers after the war. If you wanted to pursue shearing early in the year, you went to Queensland where it took place from January to May. Autumn shearing it was called. In those days lots of Kiwis came to work in Queensland and then they'd go back to New Zealand for the second half of the year.'

'Shearing,' writes Ronald Anderson, 'is something between an art and a craft.' There are two

THE PRESSER USES LOW TECH RAW MATERIALS TO STENCIL A WOOL BALE.

IT IS SO DRY THE PADDOCK IS INDISTINGUISHABLE FROM THE STATION'S AIRSTRIP.

THE PRESSER POSITIONS A STENCIL BEFORE LABELLING A BALE OF WOOL.

chief ways to learn this art-craft: shearing schools and 'barrowing'. Schools to train raw shearers and to improve learner shearers are conducted by various bodies. In this book the illustrated directions for shearing a sheep have been provided by Australian Wool Corporation shearing instructor, Roy Jerrim (see Appendix). Shearing schools have a proven track record for enhancing the skills of the young people enrolled in their programs. Nevertheless, the historical and far more common method of learning how to shear has been via barrowing. This term refers to the practice of a shearer catching a sheep just before the bell rings for the 'smoko' break and then allowing a novice to finish shearing the animal during the break period. Just as a surgeon cannot learn to perform an operation from reading a book—he must develop manual dexterity skills by 'hands-on'

Shearer Man

Shearer man like toast and butter,
Wolseley comb and Lister cutter;
You can tell those greasy shearer bastards
By the rags upon their feet:
And when they see those speed balls,
God Jesus! can they eat!
They shore wet sheep on Monday,
And they shore them wet again,
They make those poor old rousies work
In ninety points of rain.
The rouseabout he laugh and joke,
Rains come down and engine broke.

experience—so must a potential shearer get in amongst the wool. It is nor a skill learned overnight. The competent barrower would finally earn his first stand as a member of a shearing team. When that day arrived he was still a learner, a rookie in contemporary language, the slowest and least valuable member of the team. It would be many months before most novice shearers would become competent and an asset to their colleagues. Thankfully, large numbers of modest sheds were present, allowing the shearer's education to take place as a part of the overall shed operation.

Talking to two of Ivan's teams, all of the men disclosed that they started shearing by barrowing, most commonly by doing just the finer pieces of the fleece at first. Some youths have relatives already shearing and it is a natural progression for them to move into the vocation; also, rouseabouts looking to improve their lot in life make up a sizeable percentage of learners. Many men start as learners by getting paid rouseabout wages plus receiving a comb and cutter allowance.

There is a normal attrition rate in shearing estimated at about 10 per cent a year. Obviously, new shearers are required on a continuing basis to maintain the supply of workers. A recommended method in the industry calls for one out of every five stands to be staffed by learners, although in practical terms this is frequently impossible to implement.

Mulesing

A contemporary book entitled *Pulling the Wool* was written by Christine Townend, an avowed animal liberationist. *Pulling the Wool* concerns the suffering Merinos undergo while living in Australia and, in particular, it attacks the practice of mulesing sheep. The procedure is both widespread and controversial so it is fair to present both sides of the dispute.

For readers unfamiliar with mulesing Ms Townend explains:

> Mulesing is a surgical mutilation performed on sheep without the use of an anaesthetic. The skin, and (when the shears slip) the flesh ... is sliced away with a pair of shears from the tail area of the animal. This is called mulesing and the purpose is to remove the folds of skin ... and prevent moisture and urine from collecting in the skin folds around the tail.

Liberationists claim that mulesing is both vicious and barbaric. Proponents claim that it saves sheep from a fate far worse than the temporary pain of mulesing. The procedure is performed to prevent flystrike, the term used to describe the action of the common Australian blowfly. (It is also known as blowfly strike or sheep strike.) Most dangerous of the continent's flies is the greenfly, *Lucilla cuprina*, whose maggots invade the animals' skin.

In a nutshell, this is what happens. The blowfly deposits its eggs in sores or in damp or dirty wool, the kind commonly found it the hindmost region of the sheep. The larvae hatch into grubs which work their way into the sheep's skin. If neglected the animal may endure a distressing death. Blowflies have cost the Australian wool industry hundreds of millions of dollars.

It's not a minor problem. That's why many rural observers have lionised J. H. W. Mules, the South Australian grazier who in 1931 devised the operation Ms Townend abhors so passionately. The surgery has helped reduce flystrike to a problem which is now mostly manageable.

The Merino is particularly susceptible to flystrike because of its dense fleece, the wrinkly nature of the beast and its abundant body secretions. These elements plus seasonal factors mean that the blowfly can be an exceedingly dangerous pest. The Townend plan for eliminating mulesing calls for developing a new breed of sheep, returning to 'almost permanent shepherding of flocks', regular crutching and the use of coats for the animals. She believes the Merino industry should be closed down if effective control measures are not introduced. Strong words indeed.

One of the modern strategies pitted against flystrike is the jetting process, whereby a chemical to control the blowfly is sprayed under high pressure into the sheep. The area 'jetted' may be that which is crutched, the pizzle or the entire body. Weather conditions and seasonal requirements indicate the need for jetting. Ivan explains the benefits of this procedure:

> It stops the blowfly from laying eggs. Even if the fly does succeed initially, the maggots are killed. If a fly blows a freshly shorn sheep the maggots may just fall off. They usually need wool to set up residence. Jetting cuts the damage, but even today the industry loses millions of dollars to blowflies. A struck sheep is capable of returning to normal health and once again becoming a good wool producer.

A point which yields universal agreement among any shearing team is that no flyblown sheep should ever be presented to a shearer on the board. Station hands mustering the flock should detect the blown sheep before it gets to the shearer's stand. The following comments are just a few made by shearers concerning the mulesing procedure:

> The simple truth is that taking a pair of shears and cutting away the area where skin folds form is better for the sheep. It makes the skin smooth.

> Like a man it is: you do it when he's young, not when he's fifty. Away you go, knock off a bit of skin and that's exactly all there is to it.

> If she says mulesing is tortuous to the sheep, she ought to see what flies do to the sheep.

> If that woman who wrote the book had seen flyblown sheep, she'd eat her words.

Crutching

Crutching could be defined as a 'mini shear'. It's a process similar to shearing except it is done over a small area of the sheep. Pizzle wool is removed so that it doesn't get moist and attract blowflies. Wool around the sheep's anus is cleaned away so that it doesn't become daggy. Normally, the animal is 'wigged' so that it can see where it is going. This wool is then classed according to its length. The procedure is highly beneficial to the overall health of the national flock. The Australian Wool Corporation estimates that approximately 90 per cent of all sheep are crutched.

Crutching is done on a contract or non-contract basis. Ivan clarifies the terminology:

> Non-contract refers to station staff doing the crutching. Not being such a highly skilled job there are a lot of station hands who can crutch sheep, but they can't shear one. Out in the Goodooga area there are a lot of Aborigines who are not shearers but they are very good crutchers. It's a job that is really heavy on manpower; they have to catch a sheep in the pen, drag it out on the board, shear around the crutch and tail and then release it.

There has been a disagreement between pastoralists and crutching teams in the New England district. The 'New England crutch' is done on a smaller area of the sheep than is standard and so the landowners wanted to pay less for the job. The unions disagreed and Ivan could see a good deal of merit in the union claim. 'The penning up and dragging out of the sheep is the big job,' Ivan says, 'not the minimal cutting done either way on the board.' The numbers illustrated the work involved in crutching. 'An average shearer will crutch 400 sheep per day,' relates Ivan. 'A top shearer will do 800 per day. I have seen as many as 1100 sheep crutched in a single day. If we say that an average sheep weighs a hundred pounds (about 45 kilograms), then there are twenty sheep to the ton (about 1000 kilograms). Thus, an average shearer drags out twenty tons (over 20 000 kilograms) per day and a top shearer moves forty tons of sheep in one day. That is one hell of a lot of hard work.'

One shearer comments:

CRUTCHING MERINO WETHERS.

Crutching is not done as frequently now by the traditional shearing teams; it's done more often by suburban shearers. It means handling a lot of sheep. Say there are five shearers and each man does 400 sheep per day. That's 2000 sheep for the station hands to bring in and get away and then there's another 2000 that have to be organised for the next day's crutching. When the paddocks are big and the weather is hot, the sheep don't want to travel. It becomes a colossal job for the owner's men. Bringing in town shearers lets them do a day or two of crutching, have a couple of days' break to bring the next mob in, and in general there's less pressure this way.

CRUTCHING MERINO WETHERS.

7

On the BOARD

The bell rings. It's 7.30 a.m. and the start of the first of four runs of shearing that will take place today. It's the same all over Australia: a ten-hour working day broken down into eight hours of shearing with the other two hours expended in lunch and smoko breaks. Those two hours of 'down' time are completely necessary for there is no other group of workers in the nation that work harder than shearers.

SHEARER'S AIR-CONDITIONER. A PORTION OF A BEER CARTON IS ATTACHED TO THE OVERHEAD DRIVE SHAFT; AS IT ROTATES AT SPEED THE AIR COOLS THE SHEARER WORKING BELOW.

In New Zealand, especially the North Island, shearing is done for nine hours with five break periods. As Godfrey Bowen writes, 'It gives a man a great start for the day, when he walks over to breakfast with a good number of sheep behind him.' The New Zealand climate is the motivating force behind the extra hour of work. High rainfall makes the longer day unavoidable in many locales The extended periods of inactivity waiting for wet sheep to dry places the workers in a bind and means that the nine-hour day is imperative.

When the bell rings each shearer goes to his catching pen, up-ends the nearest sheep and drags it backwards to his stand. Arrangements are such that shearers either have their own pen or they share one with a pen-mate. If a catching pen is shared there are two ways to select the next sheep to be shorn. The shearer can take the animal nearest him when he goes into the pen or he can take the one which he thinks will be the best to shear. Ivan Letchford explains the second option: 'He makes his selection immediately, quickly assessing the sheep as he moves towards it. The shearer looks at its body conformation and symmetry, body wrinkles, neck folds and looks for a sheep that stands well, looks well, is not overdeveloped in the front or neck, with long staple wool, in general one that cuts and combs well.' Nobody wants to get the cobbler, which is the most difficult shearing sheep. As expected, the cobbler is often the last sheep left in the pen.

The sheep are less than keen on all this activity but the catching pen is so small that there's no problem in collecting one. The shearer drags it out to his stand, which is adjacent to the pen. The stand is the name for the area in which the shearer conducts his job. Dragging an unwilling and weighty sheep over a lengthy distance would waste time, money and effort. One quickly observes that everything is done in the shed to conserve energy and increase efficiency.

In their heyday woolsheds were enormous

structures, with those housing sixty to eighty stands not unusual. The Australian woolshed record has been claimed for both Queensland and New South Wales, where a shed in each state reputedly had 101 stands. Unfortunately for us, those structures have long ago been destroyed. Sometimes the larger sheds would have upwards of 200 men involved in shearing and allied jobs during the season. The scale of operation is now much more modest and woolsheds of four to twelve stands would be standard.

Assuming that a typical shed has eight stands, the designation goes according to custom. A draw-from-a-hat among the shearers determines who gets stand one, two, etc. Stand one is also closest to the source of shearing power. Anyone can be selected as the union rep but the most common method today is for the drawer of stand one to be the rep. He remains the union delegate for the duration of the shed. Prior to the commencement of shearing, the rep calls for a show of tickets (union membership cards) so it can be seen that each man is financial with the union.

If there are a couple of learning shearers employed on the team they would get stands seven and eight, and for a very good reason. As learners they would be slower and produce fewer fleeces so they would be positioned farthest from the wool table. A picker-up works more efficiently if the most productive shearers are near the wool table.

The shearer stands on the board, sheep to hand, ready to commence shearing. He reaches upwards, pulling a cord which engages the clutch of the shearing machinery. Power is transmitted through a 'downtube', a jointed column ending in the business end of the shearer's equipment, called simply the handpiece. Combs and cutters are fitted to the handpiece and they are responsible for, in C. E. W. Bean's quaint phrase, 'transforming the shorn bodies to the clean white of a peeled orange'. The machinery down to the handpiece is provided by the pastoralist while the handpieces themselves are supplied by the contractor. Today's shearers benefit from a rule change which allows them to provide their own handpieces, cutters and combs and earn extra income for doing so. Using their personal gear secures shearers an extra $2.50 per hundred sheep shorn, or 84 cents per hundred animals crutched. It isn't entirely clear financial gain because the men must purchase the equipment, which is fairly costly. Modern handpieces cost from $400 to $600, combs are $37 and more, while cutters cost approximately $6. Too, the shearer then loses the services of the 'expert' who formerly looked after grinding the combs and cutters. There is a considerable variance of wear on comb and cutter, depending on the status of the sheep. If they are especially dirty from grazing over dry, sandy-soiled areas, a shearer could use two combs and six to ten cutters within an hour.

Just as motorists argue about the merits of various automobiles, shearers disagree over their equipment. The two main competitors in the Australian market are the Lister and the Sunbeam companies, with a recent contender being the European firm, Heiniger.

As soon as the sheep has been thoroughly fleeced the shearer guides it towards a chute and sends it outwards, each animal arriving in its counting-out pen. The contractor goes to the counting-out pens at the beginning of each break period, tallying the number of animals shorn per shearer and then entering the figure in a tallybook. Needless to say, the contents of this little book are watched with avid interest by all the shearers.

Aiding the shearer in the shed are several other individuals. The wool industry booklet entitled *Code of Practice* furnishes today's correct titles for these workers. It includes the 'shed staff': shedhand, presser, penner-up, overseer/expert and classer, among others. There have been many terms for shed staff over the years, several of them remaining in common usage; following is a look at these people and their jobs.

The most common term for the general shedhand who does any and all jobs in the heavy-labouring class is 'rouseabout', often shortened to 'rousie' in typical Australian fashion. What the labourer is to the construction industry, the rouseabout is to the wool industry. Many shearers and pressers come from the ranks of the rouseabout who wants to better himself in the shed.

The picker-up, or fleece-oh in New Zealand, retrieves the shorn fleece and delivers it to the wool table. A good picker-up tosses the fleece onto the table in a beautiful expansive move so that when it lands it resembles a bearskin rug stretched across a floor. This board boy must be a hard worker if he or she is to be useful to the team. It was customary to have one to every six or eight shearers in the days when shearers stretched along the board toward the horizon. Today, the picker-up might also be a sweeper or a broomie, keeping the floor clean—and safe. The loose locks from the board and wool table area are kept in a separate bin. While not as valuable as the main fleece, these pieces are worth far too much to discard.

The woolroller removes the heavy sweatlocks and side wool, or skirtings, so that the classer is able to grade the wool. In general terms, the classer examines the wool for length of staple, colour of fleece and diameter of wool fibre; he also tests it for tensile strength before placing it in the appropriate bin.

ADVANCES IN THE SHEARING SHED. AN ELECTRIC FAN COOLS A SHEARER DURING CRUTCHING ON A HOT SUMMER DAY.

Being a classer long before he was a contractor, Ivan's knowledge base is far greater than most contractors who frequently have been shearers before entering the business as employers. Classing wool is more technical than one might imagine. However, under Ivan's tutelage, discussion of the subject matter is oriented to the layman's level of understanding.

'I begin by making a visual appraisal', Ivan says. 'I look at the wool and see that it's colour is good, that it's bright, that it's natural and I feel its softness. Taking a staple, I draw it out between thumb and index finger, reviewing its soundness (tensile strength). There could be a break in the staple, brought about by such things as a change in the sheep's health, by lambing or by an attack of parasites.

'The classer examines its "style", which refers to the length of the staple; its "character", which pertains to crimp or waviness; he assesses the diameter of the fibre visually and evaluates its combing quality.'

It may seem unbelievable but a good wool classer can visually gauge the diameter of a wool fibre. 'After many years', says Ivan, 'you are spot-on'.

Another yardstick appraisal is the 'size' of wool. For scores of years we have heard wool described as being 70s, 64s, 56s, etc. This relates to the spinning quality or 'fineness' of a woollen fibre. The figures refer to the number of hanks of wool which can be obtained when a pound (about 0.5 kilogram) of pure wool is spun. A hank of wool is 560 yards (about 512 metres) long and a 70s count means that 70 hanks of wool can be obtained from a pound of wool. That's a piece of wool yarn 39 200 yards (about 35 845 metres) long. Put in the practical terms of daily life, that's yarn over 22 miles (almost 36 kilometres) in length from one pound of Merino wool. It is really quite astonishing. It's not hard to see why fine Merino wool is so desirable in the textile industry. Five strands of fine Merino wool are no thicker than a single human hair.

With metrication, the old terms are becoming *passé* and wool is now measured in microns. For example, the 70s are now 19 microns; 64s are 21 microns; and 60s are 23 microns. It is surprising to learn that 17 micron wool has 16 more hanks of pure wool fibre per pound than does a 21 micron specimen. Thus, a pound of 17 micron wool is 5 miles (8 kilometres) longer than 21 micron fibre. The 17-20 micron fibre is used for sheer, light-weight, summer-style wools; quality worsted cloth is 21-22 micron; military grade cloth is 24 micron. In general, 17-24 micron wool is used in the garment trade, Merino crutchings are utilised by blanket manufacturers and the coarser British breeds commonly grown in New Zealand produce wool for carpet manufacturers.

Over one of his wife's mammoth Sunday brunches, Ivan illustrated a desirable feature of wool. Holding a staple of fresh wool he attempted to ignite it with a match. The staple singed and blackened but it did not burst into flame. 'Notice', he said, 'that wool has a safety factor over the cellulose occurring in vegetable

A MOB OF SHORN EWES IS MUSTERED FOR REMOVAL TO ANOTHER PADDOCK.

fibres which burn with a flame. Wool is 50 per cent carbon and is therefore naturally flame resistant.'

Ivan says, 'sound wool fibre has a high tensile strength, it's hard to break'. Nina Hyde corroborates Ivan's comment when she notes that 'wool can be bent 20 000 times without breaking'. The comparable figures for silk are 1800 bends before it breaks while rayon yields after only 75 bends. Warming to his subject, Ivan claims that 'wool never wears out, it only severs'. He discusses the reclaimed fibre industry where old woollen rags and by-products of mills go, and where they often add lamb's wool to put life or vitality back into the massed fibres. 'Vitality is life. Particularly notice lamb's wool. It's softer, springier, more supple, a vibrant type of fibre.'

Reviewing directions for classers, one notices that they must ensure removal of all fribs, sweats, crutchings, topknots, shanks, joints, skin pieces and dags. Ivan explains:

> Fribs are stringy pieces of wool that grow under the sheep's 'arms' and in the crutch. Sweats are waxy, greasy fibres found in the crutch. The wool cut from the sheep's crutch is called crutchings. Short bits of wool from the head area make up topknots, while shanks are matted leg pieces. Scraps of skin create difficulties for manufacturers and dyers. Dags? Well, they are

THROUGH ONE GATE, THE MOB TRAVELS A COUPLE OF KILOMETRES BEFORE FUNNELLING THROUGH A FINAL GATE AND INTO A FRESH PADDOCK.

portions of wool from over near the backside which have become infiltrated with sheep faeces; they could be pea size or pear size. In composition they're usually about 20 per cent wool and 80 per cent dung. Dags depend upon how much wool the sheep is carrying, its feeding pattern and health. Worms could give an animal the scours (diarrhoea) or switching from drought conditions to fresh green feed could cause pasturage to go straight through the sheep, dags forming as a consequence. Now when you hear that someone is a daggy old bastard, you'll know it's not a complimentary remark.

Another problem is cotted wool where the fibres become matted or felted into a mass. A common cause for this affliction is a parasite that destroys the natural lanolin—or lubricant—which keeps the wool fibres apart. If you peered through a microscope at the outline of a wool fibre it would look like the edge of a saw blade, or possibly you would say it resembled the edge of a rose leaf. What happens is that the action of the sheep walking about without its usual lubricating protection, in connection with a serrated edge on the fibre, causes the wool to become cotted. This leads to a restriction in its use to the trade and special processing is required to separate the fibres again.

Black wool is yet another problem for the wool trade, as Ivan elucidates:

> Contractors have their men trained to keep their eyes open for black wool—and there may be only a small group of fibres, the spot no bigger than a match head. It's only the shearer that can pick out this dark wool while he's shearing. He simply calls out 'black wool' and the board boy comes and destroys that small piece of wool. If there's a large amount of black wool, it can be sold. For example, some religious orders like the natural coloured fleece (for making their habits) and this creates a market for it.
>
> Black wool is usually found to be caused by one ram in a flock. You track down that ram and get rid of him. If you have black wool which gets into the processing stage where it is through the other fibres, it is bad news for the wool industry.

There is another fault whereby the wool is called a 'doggy' fleece. Here the crimps have been reduced so the fibre is straightened and irregular. It sometimes happens with older sheep. Ivan discloses doggy wool's background: 'In the past we were under the impression that doggy wool was broader in spinning quality than it appeared. Then, with the advent of scientific testing, we found that it only lacks crimp and the diameter of the fibre is unaffected. We had sections of the wool trade buying doggy wool, particularly the fine New England wools, who were laughing all the way to the bank. The fibres weren't broader at all, it was just the crimp that was affected and they were getting this wool at a discount price. Summing it up, we've got to look out for any contaminant which would reduce the value of the fleece and we must sort and label it accordingly.'

Wool is sorted and classed to be baled and then branded, the latter done traditionally by hand, just as it has been for the past 150 years. AAA classed wool is the best produced from a particular flock. 'It is the best length and the finest and brightest wool of the clip,' says Ivan. 'AAA Combing is from a medium-strong woolled sheep, of combing length, just not as good as AAA. The next grade down is AA Combing, a strong wool. AA is shorter wool, of high quality but say two inches in length instead of three inches, distinguishing it from AAA. Now that staples are measured in millimetres, I've got to say 77 mills instead of three inches. That's how you decide what wool goes in which bins. Every bale is core-tested today and so the branding on the bale is not as important as it used to be. All it is doing is distinguishing one lot from another lot so that it can be catalogued on the showroom floor; it is just distinguishing each line from the previous line.'

How much wool is obtained from each sheep? Ivan explains:

> Pre-war, it was about 7 pounds (about 3.2 kilograms) of wool carried per sheep. Now it averages about 12 pounds per head and sometimes it's possible to cut 15 or 16 pounds of wool from 'dry sheep'. There's a saying which states that the average well-bred Merino will cut one pound of wool a month. There are factors which will change that amount. In dry years, there is a

The Backblock Shearer

W. TULLY

I'm only a backblock shearer boys,
As easily can be seen,
I've shore in most of the famous sheds
On the plains of the Riverine.
I've shore in most of the famous sheds
And seen the big tallies done,
But somehow or other I don't know why
I never became a gun.

Chorus:
Hurrah my boys, my blades are set,
And I feel both fit and well.
Tomorrow you'll find me at my pen,
When the gaffer rings the bell.
With Haydon's patent thumb-guards fixed,
And both of my blades pulled back.
Tomorrow I'll go with my siding blow,
For a century or the sack.

I've opened up the windpipe straight,
I've opened behind the ear,
I've shore in every possible style
In which a man can shear.
I've studied all the cuts and drives
Of famous men I've met,
But I've never succeeded in plastering up
Those three little figures yet.

The boss walked down the board this morning,
I saw him stare at me,
I'd mastered Moran's great shoulder cut,
As he could plainly see.
But I've another surprise for him
That will give his nerves a shock,
For tomorrow I'll show him that I have mastered
Pierce's rang-tang block.

And when I succeed, as I hope to do,
Then I intend to shear,
At the Wagga demonstration,
That's held there every year.
It's there I'll lower the colours, my boys,
The colours of Mitchell and Co.
And instead of Deeming you will hear
Of Widgeegowera foe.

major reduction in wool quantity. The less food a sheep gets the less wool is produced. If parasites or other health problems affect a sheep, it will cut less wool.

Lower yielding fleeces result when a sheep has to travel long distances across dry dusty paddocks to get to water. They pick up lots of dust in the wool and we have to trim more wool away from the fleece. The wool price is based on the amount of clean scoured wool per fleece. On average, that's about 68 per cent and with skirtings the average yield is about 50 per cent. The rest is dust, dirt and vegetable material. I've seen years when it was so dirty we had to take the backs out of the fleeces as well as the edges. We have had yields as low as 27 per cent, which means that 73 per cent was dust, dirt and vegetable fault. Fleeces are generally cleaner in the north-east and Monaro areas of New South Wales. Also, finer wool is found in these higher altitudes: it's natural for wool to grow finer in the colder conditions.

As for the amount of wool cut from the country's sheep, I noticed that a recent national clip was about 800 million kilograms. That sounds about right. If the average cut was five kilos (12 pounds) per sheep and we have 160 million sheep in Australia, it squares up at 800 million kilos. Here's another yardstick saying: Power-pressed fleeces generally yield thirty bales of wool for every thousand sheep. That's 12 000 pounds (about 5400 kilograms) of wool, divided by 400 pound (180 kilogram) bales, giving thirty bales.

A couple of years or so back, the Australian Wool Corporation produced a chart which gave the average number of sheep per property. The country was divided into zones and depending on which zone was discussed the sheep numbers averaged between 2000 and 7500 per property. On these figures the number of bales would vary

ABOVE LEFT: BLACK SHEEP GENERALLY APPEAR IN THE ORDER OF ONE IN A THOUSAND. THEIR WOOL IS KEPT SEPARATELY TO AVOID CONTAMINATING NATURAL FLEECES. PIGMENTED WOOL IS EAGERLY SOUGHT BY HOME SPINNERS, COTTAGE INDUSTRIES AND RELIGIOUS SECTS WHO MAKE NATURAL, UNDYED GARMENTS.

ABOVE MIDDLE: SHEARER 'OPENING-UP THE NECK' OF A MERINO WETHER.

ABOVE RIGHT: SHEARER 'FINISHING OFF' MERINO WETHER.

from 60 to 225 per property. The average bloke on western blocks who runs about 3000 or 4000 sheep produces between 90 and 120 bales of wool. Of course, the figures come apart in Victoria where we've got poultry farmers, orchardists and hobby farmers with a few sheep eating weeds around fruit trees. A chap producing five bales becomes a statistic and he throws off the average. A few days ago, a representative for one of the wool firms told me that the present average clip out of Newcastle is eighty bales.

When the classer has finished his work the wool finds temporary storage in various bins from where the presser obtains it. Historically, records indicate that wool was 'pressed' on John Macarthur's property in the early 1800s. Initial models were simply screw presses, while later in the century rack-and-pinion versions appeared. Surprisingly, the hydraulic press was not ushered in until the 1960s. Ivan believes a portable hydraulic press he invented in the 1960s was a first for this industry. 'I never bothered to patent it', Ivan says, 'although I sure have had many long years of faithful service from it.'

The presser packs the wool into bales ready for transporting. Today's bales are square-ended with an overall rectangular shape. The bales measure 685 millimetres on each end surface and are 1370 millimetres high. Where the bale came from is unknown but it was possibly formulated on the basis of the amount of wool borne by a camel. These ships of the desert had been used from early days to convey wool in outback regions.

The presser is a man who frequently starts off as a general hand and then takes on pressing to upgrade his station in the shed; as well, it is a better paid position. In days of old the presser was always a big strong individual because he did the hardest work in the shed. Now the mechanical press enables anyone to take on the job. Nevertheless, pressing is not a job for shirkers. In fact, there's no room for bludgers anywhere in a shearing shed.

SKIRTED AND ROLLED FLEECES WAITING TO BE CLASSED.

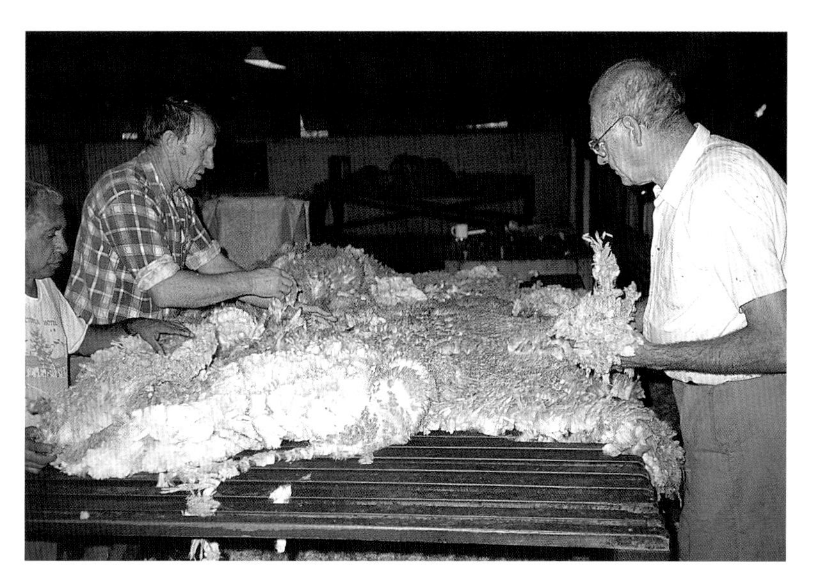

SKIRTING A FLEECE TO REMOVE SWEAT LOCKS AND VEGETABLE FAULTS FROM THE EDGE OF THE FLEECE.

The presser's contract is different from those of the other shedhands. He signs his agreement before starting the job, nominating whether he wishes to be paid by the bale or kilogram. If he was to work with dense heavy wool he might choose to be paid by the amount of weight pressed; on the other hand if he was to deal with a superfine specialty clip where they don't want heavy bales, it would be to his advantage to be paid by the number of bales pressed.

Bales are packed with a minimum of 110 kilograms and a maximum of 204 kilograms (240 to 450 pounds) of wool. Ivan explains that the most common method today is to aim for about 400 pounds (181 kilograms). 'Wool is sold by the bale, freighted by the bale, warehoused by the bale and there's a saving in testing and woolpack costs for heavier bales. So it's to the advantage of the grower to get as much as he can into the bale from the standpoint of cost-cutting.'

Bale weight has gone up now that hand presses have been superseded by hydraulic presses. Earlier in the century the old hands liked to keep a bale around 300 pounds (about 136 kilograms) in weight. The argument was that if the wool was pressed too tightly it wouldn't open up properly on the showroom floor. Ivan expands: 'The wool didn't "bloom" correctly, it took on a dull, lifeless appearance. With extra superfine wool, packing lightly gave excellent bloom, allowing the wool to open up nicely on the showroom floor.' Modern technique calls for the filled bale to be sealed with a bag fastener. What used to happen was that the presser sewed bales shut with needle and twine. A cap went with every woolpack, the entire parcel made of jute. When the wool was pressed, the presser placed a cap on top and used pins or skewers to secure the cap to the jute pack. The caps were 27 by 27 inches (about 69 by 69 centimetres); cut in half they made perfect raw material for 'bag boots', a form of footwear once popular among shearers.

Today the baling material is no longer jute; it is a synthetic material called polypack for average wools, while nylon is used for superfine wool. Ivan discusses synthetic fibre: 'jute wasn't injurious to the wool but the polypack has the potential to damage its contents. When the bales are handled with hooks and other equipment the synthetic fibre can 'shatter' if it is driven inwards, injuring the wool. Now, the usual way to get vegetable matter out of greasy wool is to run it through a very weak solution of hydrochloric or sulphuric acid. Wool is relatively unaffected by acids but it is affected by alkalis. After scouring, the wool is put in dryers where the heat turns the vegetable matter into a powder or ash, which is removed completely. This standard process doesn't remove the polypack material.'

The final step done by the shearing team is the presser's job of branding the bales of wool. While this is much less important now, it is still conducted in a thoroughly professional manner. They use a variety of perplexing abbreviations. Nonetheless, all should be readily decipherable. A sampling: PCS (pieces); STN (stain); LKS (locks); CRT (crutchings); DAG (dags); SKN (skin); BLS (bellies); COT (cotted); and COL (unscourable coloured wool).

Pressers are not only hard working, they are subject to injury from the powerful press. Joan Palmer recounts one of the most macabre stories to have come out of the shearing shed. Many years ago an Aboriginal girl was employed as a picker-up on Tambo station in Queensland. She disappeared after lunch one day, never to be seen again—alive, that is. The mystery was solved in England several months later when her body was discovered inside a bale of wool. It was surmised that she had enjoyed a pleasant nap at lunchtime but then the sudden onset of noise as the shed started up and massed wool came tumbling down on her obscured her cries for help.

An important member of the shearing team is identified as the 'expert'. He is the person in the shed who is responsible for the machinery operating smoothly. More significantly, he is known for the job of grinding the shearers' combs and cutters. When bales were used in the last century each man owned a stone, and if it was of high quality the stone became a prized possession. Stones were carried from shed to shed inside the swag. The stone was used to maintain a keen edge on the blades so that they would seem to glide through the wool with very little jawing motion. When sheds were huge and staffed by large numbers of shearers, the expert's task was full-time. With today's smaller teams experting becomes a part-time job for one man. In Ivan's case he can usually look after experting duties during smoko breaks and at lunchtime. He has gone to great lengths to explain how necessary it is to grind the combs with exactitude, not grinding them down too much. Veteran shearers frequently credit the comb with being the most significant portion of the gear, declaring that the status of the comb creates a solid basis for competent shearing.

From the ring of the first bell at 7.30 a.m. until the final bell of the day sounds at 5.30 p.m., shearing is flat-out hard work. Because of the internecine struggles between early shearers and landowners, because of union/non-union strife over the wide-comb issue which has been well remembered by the public, and because of the knockabout qualities of some shearers, the shearing industry has not always enjoyed the most salubrious of reputations. If the Australian people realised what good value and enhanced prosperity came to the nation from their toil, a Shearers' Day or Shearers' Week would promptly be proclaimed.

8

CALLUSES *or* COMPO

We were really pissed on a Saturday afternoon in Walgett and we were driving back to the shed when he fell off the back of the truck. We saw him in the mirror right away. We backed up and said:

'Are you all right, mate?'

'Yeah,' he said, 'I'm a bit sore but don't say nothing about it. Monday morning I'll just fall down the stairs and go on compo.'

Sheareres are usually a very healthy group of people. They have to be if they are going to work exceptionally hard on a day-to-day basis. This chapter concerns the health of shearers and it has been encased in the words 'calluses or compo'. The calluses portion of the heading is metaphorical and pertains to the thickened, horny skin which results from hard labour. As for real calluses, the shearer won't have the hardened hands one expects from such strenuous endeavours because he works all day with his hands in lanolin. The compo portion of the title refers to those instances when a worker isn't fair dinkum with his mates and compo is used to secure undeserved personal gain.

TOOLS OF THE TRADE. ON A SHELF BEHIND THE SHEARER RESTS A RANGE OF ITEMS NECESSARY TO HIS WORK. A FOLDED CONTRACT HANGS ON A NAIL. THERE'S A NEEDLE FOR SEWING UP A CUT SHEEP. A WASH-UP BRUSH WITH ITS BLADE ATTACHMENT IS USED FOR REMOVING THE NATURAL YOLK FROM COMB AND CUTTERS. THE SHEARER'S OIL POT LUBRICATES THE HANDPIECE. THE PALM-HELD SCREWDRIVER ALLOWS FOR CHANGING COMB AND CUTTERS. THE WALLET HAS SIX POCKETS FOR COMBS OF VARYING THICKNESS. THE CUTTERS ARE STRUNG ON A PIECE OF WIRE AND USED IN ROTATION TO PROVIDE EVEN WEAR. LAST BUT NOT LEAST, THERE IS AN ENAMELLED DRINKING MUG FOR TEA.

Three veteran shearers explain the use of compo:

Can you shear for thirty years without serious health problems? Sure, you can shear for forty years. Some blokes shear for five years and go for a 'lump sum' and some shear for forty-five years without any trouble.

I've seen blokes shear for five or six years who wouldn't shear enough to make a stew. They go for the lump sum and bang, they've got $70 000. Then you get a true and honest bloke that shears for thirty-five or forty years who goes for a lump sum and honest to goodness, he'll get $10 000 or $11 000. It depends on whether or not you are going for an Oscar award. Are you going to be a Richard Burton in the acting game or someone else?

That's why insurance is so hard for the shearers. You'll have the honest bloke putting up with all his pains for thirty or forty years and you'll have the bloke who's been in it for five years. He'll lie down on his back for two or the years and he'll pick up seventy or eighty grand. As soon as it's all finished he's back working again.

MANY SHEARERS WEAR 'BAG BOOTS' WHICH WERE ORIGINALLY HAND MADE FROM JUTE WOOL PACKS AND ARE NOW COMMERCIALLY MANUFACTURED. THEY ARE CONSTRUCTED WITHOUT HEELS AND ARE VERY COMFORTABLE TO WEAR ON THE BOARD.

Just as with other lines of human endeavour, there are shearers and there are shearers. Historically, there were differences in the competence of shearers but there was no elaborate social welfare system

SHEARING IS EXTREMELY HARD ON THE BACK. THIS PAIR HAVE ADOPTED AN INCREASINGLY POPULAR SUPPORT MECHANISM WHICH LESSENS LOW BACK STRAIN.

One of the Has-Beens

I'm one of the has-beens, a shearer I mean,
I once was a ringer and used to shear clean;
I could make the wool roll off easy,
Like soil from the plough,
But you may not believe me, because I can't do it now.

Chorus:
I'm as awkward as a new chum,
And used to the frown,
That the boss often shows me,
Saying 'keep them blades down'.

I've shorn with Pat Hogan, Bill Bright and Jack Gunn,
Charlie Fergus, Tommy Layton and the great roarin' Dunn.
They brought from the Lachlan
The best they could find,
But not one among them could leave me behind.

Still it's no use complaining, I'll never say die,
Though the days of fast shearing for me have gone by.
I'll just take the world nice and easy,
Shear slowly and clean,
And I merely have told you just what I have been.

available for the bludgers to manipulate.

In those days there were workers and shirkers, with the latter group having it far more difficult than today. There's no doubt that pre-1900 economic conditions created undeserved hardships for many people. On the other hand, some of the shirkers were drawn into line simply through economic necessity.

Nineteenth-century shearers worked, endured, did it the hard way. One of the most common ailments affecting the men was known as 'swollen wrist'. It was a complaint each man knew about, quite often from personal experience. Swollen wrist seems to be related to the RSI epidemic which swept the nation in the first half of the 1980s. Continued movement in the wrist area plus the effort of opening and closing the shears resulted in frequent strains in the wrist ligaments. Almost as universal were the back problems which resulted from long hours of near-constant stooping. Duke Tritton recalls the difficulties at Charlton, an early-season shed: 'It was not unusual to see a man making his way on hands and knees to his bunk. No-one offered assistance as it was a point of honour to be able to reach one's bunk under one's own steam. And the man would be at his pen next morning.'

Since shearers continue to work in an exaggeratedly bent manner, they continue to have far more than the usual number of back problems. Sometimes men can only regain the upright posture at the end of a run by placing their hands on their knees and 'walking' up their legs until they stand erect. Moreover, shearers are no strangers to orthopaedic surgeons' scalpels. Still others wear back support belts to provide relief to an aching lower back.

Here's former shearer, Dennis Ryan, commenting on the way some men shear:

> It's where you are standing that does everything. People who shear with their hands are fools, people who shear with their feet last a long while. I go to some of these shearing competitions and I see blokes in the most ungainly and disgraceful postures trying to shear a sheep. Half of the time it looks like the sheep is shearing them. They get the wool off quick enough, but what they are doing to their discs and joints I'm damned if I know.

Another common affliction of the 1800s was known as 'yolk boils'. The term 'yolk' refers to the naturally greasy material produced by the sheep's skin glands; Alfred Hawkesworth simply says it is 'sheep's grease'. The amount of yolk varies, and when it is heavy a shearer will complain of the yolk building up on his combs. Yolk can act as a carrier for micro-organisms which can gain entry to the shearer through burr or thistle abrasions, minor scratches, etc. Common on forearms, knees and thighs, serious yolk boils can scar quite badly. In modern times the men have become aware of the importance of personal cleanliness in maintaining health. 'Shearers and shedhands are extremely clean people,' says Ivan Letchford. 'They shower after work each day and as a preventive measure they wash their dungarees on a daily basis. Yolk boils occur with dirty clothing and everyone wants to avoid them.'

The fascinating subject of shearing wet sheep has been a contentious issue in the sheds for well over a hundred years. During the last century, shearers declared that shearing wet sheep led to arthritis, rheumatics, non-specific internal complaints, headaches, sore eyes and skin complaints. Here it is near the end of the twentieth century and they are saying the same thing. It's not proven one way or the other, but the men believe it so strongly that provision has been made so they don't have to work if they think the wool is wet.

A medical inquiry was launched into the wet-sheep question in 1908. The physician heading the investigation talked to more than a thousand shearers, many fearful of wet sheep. Only 65 voiced specific problems which had resulted from shearing wet animals. Of the 65 there were 30 with knee complaints, but 15 of these were judged due to an age-related rheumatoid arthritis. Another 7 had injured synovial tissue, 7 more had gonorrhoeal arthritis and 3 had a history of rheumatic fever. The doctor declared that shearing wet sheep didn't cause knee disorders but could aggravate existing problems. Interestingly, it was found that the temperature of sheep is 4°F (approximately 2.2°C) higher than human temperature. The air around a shearer will increase in humidity if the wool is moist, creating a decrease in evaporation with consequent overheating and a possible chilling effect to follow.

Several decades ago a conference representing pastoralists, union and federal government adopted Richardson's and Gillespie's Wet Wool Detector to measure moisture in the fleeces. At this same time—the early 1900s—C. E. W. Bean talked to a veteran of forty years' shearing. The man said he'd seen wet sheep shorn frequently, especially near cutout when the men want to get to the next job: 'The sheep are never wet then.' Conversely, when the shearing was just starting, or they were near a pub, the sheep got wet very quickly. Duke Tritton told an amusing anecdote about an old blade shearer who said the sheep were much too wet to shear. The rest of the team disagreed and so they continued shearing. When he cut a restless animal the boss warned him to be careful because the sheep was covered in blood. Boss and all guffawed at his reply: 'That ain't sheep's blood. That's outa the frogs in the wool.'

Suppose a situation developed where the sheep could be considered wet. Such an instance would be dealt with through a 'wet vote', a procedure so entrenched in practice that it is placed in the federal Pastoral Industry Award. The vote is conducted by the union rep after the men shear two sheep each. Two tickets—'wet' and 'dry'—are issued; the poll is not open but secret, and the ballot box is then brought to the contractor. Together, the contractor and the rep count the result. Naturally, the majority rules. If it is a drawn ballot work continues. Shearers who voted wet have the right to take the day off if they so desire—without pay of course. The award states that a second vote on the same day may be requested by the employer. The sheep would be turned out and two hours later the question would be re-balloted. However, the union in New South Wales says there's to be but one vote per day and so that's how many votes there are.

Ivan says the shearers can have other votes, for example, if they think the sheep are too full or if it is too hot. And the men won't shear 'off the dog', which means that they won't go ahead if the dogs have just chased the sheep into the yard. If sheep are too full of tucker or water and they are put on the board, pressure is put on them in shearing; this causes them to kick and struggle more actively. 'It's a logical rule that all dry sheep must be yarded for four hours before being shorn,' says Ivan.

Living conditions for a shearing team have improved immeasurably over the years. Nonetheless, benefits gained were frequently won grudgingly. For example, it is quite astonishing that mattresses for beds in the shearers' quarters were only secured in 1947! Until that point the men had palliasses, which are mattress covers filled with straw. 'The owner would give you a bale of hay to fill your palliasse', Ivan recalls. 'Some of the miserable cockies wouldn't even buy good straw. Instead, they'd cut meadow hay to give us and often it was full of burrs and thistles.'

Present-day conditions are seen by grazier Ron Taylor as quite acceptable. 'The accommodation was definitely not good at one time, although there's not much wrong with it today. There's a cook provided, a meal every two hours, refrigeration, hot and cold showers. You can't wish for too much better.'

Ivan discusses owners and their importance to a successful shed:

An owner must be on side or he can place obstacles in the team's way; in fact, he can make life hell or easy for the men. His attitude is responsible for many of the feelings—good or bad—in a team. Unions had to come about. Early owners 'raddled' and gave the shearers atrocious accommodation. If a shearer has crook facilities, a poor mattress, toilet and shower, 'bad' water and, say, no wood cutter, he is disadvantaged.

In the shed the sheep need to be presented in good order. If it is a tidy, even, well-managed flock, well classed with the wrinkly sheep removed, with the sheep dressed for flies and having no parasites in them, it is much easier all round.

Keeping the men healthy is partially keeping the men happy. Harmony is most important to a shearing team. You must keep a good cook and watch out for excessive drinking. It's best if the men stick to a regular routine, only a few beers after knock-off, dinner at 6.30 p.m., no real drinking after that, and getting to bed nice and early. They work very hard and need the rest. This suburban shearing where a bloke drives up to a hundred miles (about 160 kilometres) in one direction after a 7.30 a.m. to 5.30 p.m. working day is both foolish and unnatural.

HEATED WATER FOR THIS SHOWER BLOCK COMES FROM A 'DONKEY BOILER' MADE FROM TWO 44-GALLON DRUMS.

Perhaps this is the time to quote a shearer known as 'Banger'. He says that shearers are just like babies:

1. They need a feed every two hours.
2. They want a bottle every night.
3. And, they whine all the time.

Safety in the shed is often related to the presence of vegetable matter in the animal's fleece. The substance can cause irritation to the shearer in a direct fashion or it can cause the shears to work erratically, indirectly injuring a shearer as the machinery runs amuck. From a historical standpoint, the prickly pear was a real hindrance to shearers and graziers alike.

It is a cactus variety which came to Australia at an indeterminate time, the source disputed. Its flat leaf has fine barbs about six millimetres long which are nearly impossible to remove. In the skin of a shearer's legs or upper extremities they would fester and create nasty sores. If men were employed in an area infested with prickly pear, the pastoralist provided chaps to wear over jeans to ward off the barbs. The problem steadily worsened until 1920 when the prickly pear's advances became alarming; in some instances graziers were forced to abandon their land.

Deliverance disembarked in 1925 in the form of a caterpillar from South America, *Cactoblastis cactorum*. Success was rapid, the ravenous cactoblastis destroyed 26 million hectares of prickly pear in less than a decade.

Tinkering with imported flora and fauna in Australia has been commonplace but of dubious benefit. Morning glory, lantana and Paterson's curse are examples of plant life which have afflicted the countryside. In the animal kingdom the European rabbit and the Queensland cane toad are two imports continuing to upset local ecology. Just the same, one must be even-handed and acknowledge that without the importation of Merinos, we would all be less prosperous.

The most common burr is the trefoil clover. 'They are not a major problem for a shearer,' Ivan says, 'they just make it a bit difficult for him to break

SHEEP ENJOYING THE ONLY SHADE AVAILABLE TO ESCAPE THE MIDDAY SUN.

out on the first leg and coming up the neck.' As for grass seed, Ivan declares: 'It's more of an irritation to the sheep than to the shearer.'

Not so innocuous is the Bathurst burr. It is compulsory for a shearing machine to be equipped with a safety clutch. Should the operator come in contact with an obstruction such as a burr, a piece of wire or a hard twig, the clutch disengages the comb and cutters, pulling the mechanism into neutral. The Bathurst burr is described by Ivan as, 'extremely durable, rock hard with thick seeds about ⅜ by ⅛ inch (9 by 3 millimetres). They can get between the comb and cutters, 'locking' a machine, throwing it out of the shearer's hand. Since the cutter travels at 3000 oscillations per minute, it can be a dangerous tool on a shed floor.'

A TYPICAL CANVAS WATERBAG, FIFTEEN INCHES IN LENGTH. ON AN ORDINARY DAY A SHEARER WILL DRINK THE CONTENTS OF ONE WATER BAG WHILE ON AN ESPECIALLY HOT DAY HE MAY CONSUME TWO. IVAN NOTES, 'THESE EVAPORATIVE COOLERS USED TO COST ONE SHILLING PER INCH AND NOW WATER BAGS COST ONE DOLLAR PER INCH.'

Shearers must dress for the conditions in which they work. If one forms a mental picture of a shearer, it is generally of a man in a blue singlet and jeans. Upper-body wear includes cotton, flannel or denim and the tails on singlets and shirts need to be extra long. This is so that the bent-over worker doesn't have a gap between shirt bottom and belt, thus preventing chills from occurring in the exposed lower back area.

For years the sleeveless styled T-shirt/singlet has been called a 'Jacky Howe', after the famous shearer. According to Patsy Adam-Smith, someone felt constrained by their sleeves and simply tore them off, saying, 'I'll make a Jacky Howe of it.' The ironic part is that his family affirm Howe never wore sleeveless shirts; in fact his wife, a dressmaker, made his sleeves snug fitting so the sheep's hooves wouldn't tear them.

Another item of wearing apparel is inextricably linked to the shearer. Bag boots are as closely identified with shearers as elastic-sided boots are with graziers. Bag boots look like elementary moccasins, are heel-less, and earned their name from their raw material, the jute woolpack. In the previous century, bag boots were cut out with the blade shears carried by the shearers. 'Crafting them by hand's a dying art, not surprisingly because there are hardly any jute packs left today', Ivan states. So now the shearer purchases commercial bag boots.

Two shearers comment:

> When woolpacks were made of jute the bag boots were made from jute. I made me own boots then. Now that the woolpack is made of nylon I can't make me own. These ones I have on are exactly the same, but they are made out of felt.

> I came up through a hard school. I never learned how to make boots although my dad—who died a couple of years ago, aged eighty-odd years—made his. Probably he couldn't afford boots and that was the start of how they made them. Years ago if you couldn't afford boots you went barefoot to shear.

Exactly what are the advantages of the bag boots which make them so popular? It's explained in direct unadorned testimonial from the shearers: 'With bag boots you don't get backache, you don't slip on the shed floor and they are comfortable to wear.' As Bowen says, 'All shearers should use this type of footwear in preference to sandshoes, slippers, shoes, or bare feet, all of which I have seen used.' In fact, Bowen gives a recipe for making bag boots, which are called moccasins in New Zealand. Here's the list of ingredients: 'One good sack, five lengths of twine, a packing needle, and blade shears.'

Sex

In *Pastoral Properties of Australia* Peter Taylor tells a story about Otway Falkiner, one of the famous Falkiners of Boonoke. One of his station hands had to go to court, charged with an unnatural act against a sheep. In court, Otway supposedly was giving the chap a reference but in actuality he was proclaiming the wondrous quality of the Falkiner sheep. Tired of the self-praise, the magistrate reminded Otway that he was in court to provide a reference for the station hand. 'Oh that,' said Falkiner. 'Well, I give each of my station hands two sheep a month. They can either eat them or stuff them and I don't care which.'

After telling Ivan this story, I reminded him of the numerous off-colour tales about men and sheep and asked him if he had ever heard of any such incidents in all his years in the bush.

'No, I've never known of any instances of men involved sexually with sheep. I've only known of a single case and that was where a blacksmith was accused of doing a similar act upon a mare. He was supposed to have been standing on a 44 gallon drum doing the act. His evidence in court was that he was standing on the drum to get down from the rafters the steel from which they made the horseshoes. The owner of the horse gave evidence that she was hard enough to shoe, let alone attempt to have sex with her. She would try to kick your head off while you were trying to nail shoes on her hooves. How could you perform any other act on her? So that was that, the case was dismissed.'

The last word relates not to sheep but to shearers. One of the team declared, 'That's not true about shearers having small families. I've had eleven kids and I wasn't too far away and I wasn't too tired either.'

In mentioning tiredness and distance the shearer was referring to the aphorism which explains the impoverished conjugal relations supposedly suffered by shearing couples. The aphorism is self-explanatory: 'Friday too tired, Saturday too drunk and Sunday too far away.' The last four words made up the title of a celebrated movie released in 1975 which starred Jack Thompson.

John Robert (Jack) Howe

(1861?-1920)

It was not enough for a man to know how to shear. To achieve recognition he needed to become reasonably fast and to shear cleanly. If he could obtain regular work during the shearing season he would be recognised as a man of some standing. If he rang a shed—that is, if he shore more sheep than anyone else his name would become well known; and if he consistently rang big sheds he would become a legend in his own lifetime.

So wrote John Merritt in a passage that could have been an epitaph for Jack Howe. There is considerable evidence for describing Jack Howe as the greatest shearer Australia has ever seen. Patsy

JOHN HOWE

FRESHLY CRUTCHED AND WIGGED SHEEP BEING COUNTED OUT.

Adam-Smith called Howe 'The Bradman of the Shearing Board'.

He is famous for two records, both of which have not been broken since the time they were set in 1892. Record number one was the shearing of 321 sheep (nine-month weaners) in a single working day with *blade* shears. This record has been surpassed but never by blades, only with machine shears and then not until 1930. Record number two stands on the books irrespective of hand or machine shearing. In forty-four hours, one standard working week, Howe shore 1437 sheep.

Jack Howe was a Queenslander, the son of a physically unremarkable mother and an acrobatic circus-performing father. Young Howe was to become a bear of a man with a chest measurement of 127 centimetres, 70 centimetre thighs, biceps of 43 centimetres in circumference, and forearms of 37 centimetres. He weighed 114 kilograms and was a superb athlete. Hector Holthouse states that during Jack's youth, much first-rate shearing in the Darling Downs area was done by Chinese workers. Apparently, these highly skilled men taught young Jack Howe many of the finer points of proficient shearing. Obviously, he learned well. All during his shearing career he won awards for 'best and cleanest' shearing results. It was customary in those days for graziers to pay a premium at cut-out for the individual who did the most meticulous job of removing fleeces.

A STARK RESULT OF DROUGHT.

He continued in this manner until goaded by a vainglorious individual who has been described elsewhere as a 'flash cove'. Howe out-gunned the gun shearer so badly that he melted away from the

TRAVELLING STOCK OF SEVERAL THOUSAND SHEEP ALONG A STOCK ROUTE. DEPENDING ON SEASONAL CONDITIONS, SUCH A MOB MAY BE ON THE ROAD FOR TWELVE TO TWENTY-FOUR MONTHS.

shed, sinking into oblivion. As with gunfighters in America's Wild West, there were plenty of others willing to try him out. None was successful and Howe rang each and every shed in which he worked.

In 1892, Sheffield Shield cricket began and gold was discovered at Coolgardie, Western Australia. On 10 October in that year, Howe shore his record 321 sheep in seven hours and forty minutes. (He could have done more but for his involvement in horseplay with some other men.) This all took place at Alice Downs in the Barcoo area of Queensland. Howe also was masterful with the newly adopted machine shears, winning gold medals in 1892 for shearing with both blades and machinery.

The shearer with a hand 'the size of a small tennis racket' (*Australian Dictionary of Biography*, Vol. 9) was weary. In 1900 he gave away shearing and became a publican. He purchased the Universal Hotel in Blackall. In 1902, he acquired the Barcoo Hotel and then he repurchased the Universal Hotel in 1907. Later on in his comparatively short life he bought Sumnervale and Shamrock Park, two properties in the Blackall area.

Howe was a strong supporter of unionism and the Australian Labor Party; in 1909 he became president of the Blackall Workers' Political Organisation. In 1919, when he moved from Blackall, the townspeople threw their biggest party ever for the community's favourite son. It is said that Howe's health was broken by this time. For whatever reason, he passed away in 1920, leaving behind his wife, two daughters and six sons.

There are individuals who say that Howe wouldn't have set his records with today's heavier fleeced sheep. It's true that yields have increased twofold and more since the late nineteenth century. It's also true that those 1800s sheep came equipped with wrinkles which have been bred out of today's animals. Howe did his shearing in the shed, not with the carefully prepared sheep and favourable conditions that mark today's shearing contests. Fair's fair. Jack Howe was not only a Bradman, he was a Kingsford Smith and a Dawn Fraser for good measure.

9

WOOLSHEDS

Oh, what marvellous structures are the woolsheds of Australia. The history of the nation can be traced through these buildings. Moreover, woolsheds are the closest thing to an old jumper that one can think of. They are well suited to their purpose; they are comfortable; they are quite individual and they normally last many more years than anyone ever expects. Is it more than coincidental that the jumper began its manufactured life in the very structure to which it is now compared? Affectionate feelings directed toward woolsheds are limited to those buildings which are constructed of timber. It would be quite impossible to develop any kind of similar feelings for something fabricated of galvanised iron. Harry Sowden created a pleasing mind-picture when he described woolsheds as 'vernacular architecture'.

Entering a quiet shed is much like entering an empty, ageing church. The silence induces an autonomic response: there is immediate relaxation; a feeling of well-being; an innate curiosity about all the people who have been there before; a sense of history that is unerringly absent from the forty-second floor of a concrete and glass tower.

It is a part of that same feeling we all get in a forest or along a deserted seashore. There is an affinity for, or perhaps one can say a direct connection with, the natural. It must be obvious to most of us that lumping between 3 and 10 million people into a modern city for extended periods of time is inimical to emotional and physical well-being. That's why anyone who's ever been in an old-fashioned woolshed looks forward to a return visit.

The upper Hunter shed pictured (page 13) has been torn down. In another place, a marvellous brick edifice of two stories has been built to take the shed's place. While the new structure might be quite impressive, it leaves the observer with no greater emotional response than that felt walking around inside Bargain Billy's Warehouse. The former shed was something profound: the surrounding grounds pounded flat by many thousands of cloven hoofs, the timber so smoothly worn, the air dappled with dust, the innumerable tales left hanging in the still atmosphere, put there by shearers yarning well before Federation. Such sheds should be protected by the National Trust. They are a form of national treasure.

Picture another shed past the Narran River on the way to Goodooga, just south of the Queensland border. It's very cold on these midwinter nights which are as clear as a church bell. The presser is on his Pat Malone, working all sorts of odd hours trying to keep up with the shearers. The venerable shed sits out on the flat terrain, stars are visible throughout the night sky and the warmest yellow glow emanates from the shed windows. Half-frozen on icy July nights, an observer would expect 'close encounters' to be enacted momentarily. Although this scene took place a few years ago, the subconscious can instantly call up a crisp re-run, even today. Can you possibly imagine such a reaction ensuing from watching the Megacity

The Shearer's Dream

HENRY LAWSON

I dreamt I shore in a shearin'-shed, and it was a dream of joy,
For every one of the rouseabouts was a girl dressed up as a boy—
Dressed up like a page in a pantomime, and the prettiest ever seen—
They had flaxen hair, they had coal-black hair—and every shade between.

There was short, plump girls, there was tall, slim girls, and the handsomest ever seen—
They was four-foot-five, they was six-foot high, and every height between.

The shed was cooled by electric fans that was over every shoot;
The pens was of polished ma-ho-gany, And everything else to suit;
The huts had springs to the mattresses, and the tucker was simply grand,
And every night by the billabong we danced to a German band.

Our pay was the wool off the jumbuck's back, so we shore till all was blue—
The sheep was washed afore they were shore (and the rams was scented too);
And we all of us wept when the shed cut out, in spite of the long, hot days,
For every hour them girls waltzed in with whisky and beer on tr-a-a-a-ys!

There was three of them girls to every chap, and as jealous as they could be—
There was three of them girls to every chap, and six of 'em picked on me;
We was draftin' 'em out for the homeward track and sharin' 'em round like steam,
When I woke with my head in the blazin' sun to find 'twas a shearer's dream.

BIG SOUTH AUSTRALIAN-BRED WETHERS MAKE THE SHEARER'S WORK EVEN MORE ARDUOUS.

Highrise in a night sky? Remember, if you have never seen an old shed, seek out an opportunity to visit one when it is busy with shearing and again when it is sitting empty, filled with the sounds of history.

The earliest woolsheds were barely buildings and then only in the crudest sense of the word. Posts were set in the ground and a bough roof erected to provide shade and some protection from the elements. Fabric spread over the ground acted as a rudimentary floor. Walls were nonexistent. Such an ephemeral arrangement could not last, particularly as the sheep began a rapid expansion in their new homeland.

Bush builders in the 1840s were adept with adze and axe. In addition saws, augers, plumb-bobs, chalk-lines and chisels teamed with spades, pick-axes, square rulers and short rulers to make up the complete builder's tool kit. These few tools did an adequate job of creating a basic building but the very nature of post and beam construction created difficulties for the areas within.

A totally new system of building was needed and came to be developed in America. It took the

continent by storm and soon arrived on Antipodean shores. Balloon framing received almost immediate acceptance. In *Australian Woolsheds* Harry Sowden explains the benefits of balloon framing:

> Machine-cut timber, wire nails and sheet iron, all the outcome of a takeover by machines produced by the Industrial Revolution, were the necessary basic materials. The previous construction methods, using heavy oversized logs, intricately jointed and locked by timber pegs or fixed by handmade nails, gave way to the rise of precision-cut rectangular sectional timbers that only required butt joining and banging together with factory-produced nails of standard lengths and shank sizes to provide quick and economic shelter.

Then, a few years down the road, sheet iron came to be used as a wall-cladding material. Corrugated iron had been patented by a Londoner in 1829. It was a natural for use in Australia. Cargo ships took wool to Britain and returned with corrugated iron; the

IN VARYING TERRAIN, THOUSANDS OF WOOLSHEDS ARE SCATTERED OVER THE COUNTRYSIDE

VENERABLE TWO-BOX FERRIER'S PRESS MANUFACTURED BY HUMBLE AND SONS, GEELONG IN THE LATE 1800S AND STILL IN USE AT HISTORIC BELLTREES NEAR SCONE, NEW SOUTH WALES. DURING A SINGLE SEASON BEFORE THE TURN OF THE CENTURY 180 000 SHEEP WERE SHORN AT BELLTREES.

material made excellent ballast and was easy to transport. The newly developed galvanising process ensured the acceptance of corrugated iron in roofing, tanks, drums, gutters and downpipes. Despite its wide usage, the material was quite unattractive, drawing the ire of some aesthetically-minded citizens. For example, Henry Lawson's remarks insinuated that greed was the stimulus for corrugated iron's introduction to Australia. No matter. Landowners rejoiced in it for it was so much lighter, more workable and more readily transported than the heavy timber used previously. An English company had taken out a patent to manufacture corrugated iron in the colony of Victoria in 1858.

Nature created a solution for another shearing shed problem. Timbers rotted when the shed was built on the ground. Elevating the building via lengthier piers kept it off the ground and increased the circulation of air around the shed. Rotting was halted and—*voilà*!—the room beneath the shed provided holding pens for many hundreds of sheep. With the animals protected from rainfall there were fewer days lost due to wet votes.

Two further building materials are worthy of brief mention. Brick began to be used in building settlers' homesteads, but it was such a durable and prestigious substance that it was generally shunned in the coarser, 'She'll be right, mate' construction of the woolshed. When brick was used the labour was frequently done by convicts. In western Victoria a variety of basalt rock, called bluestone, came to be used in sheds. There was minimal timber available while bluestone lay scattered over the countryside, waiting for laborious but ready collection.

Overseas, in 1870 Vladimir Ilyich Lenin was born, Charles Dickens died, W. G. Grace and his brothers founded the Gloucester Cricket Club and John D. Rockefeller established the Standard Oil Company. In Australia, in 1870 Western Australia was granted representative government and the woolshed had begun to take on a new role. It became the focus for a group of buildings. Nearby were huts for the shearers and, in those days, separate huts for the shedhands. During this period, shearers were known as the 'princes of labour'. Also a part of the small complex was a mess building; the men had to be fed well if they were to be sustained in their back-breaking labour.

The timber used in foundation, framework and—where utilised—cladding of the woolshed usually was comprised of locally grown varieties. Stringy-bark and ironbark were popular for slab construction and for shingles. When machine shears were introduced, a 12 by 3 inch (305 by 76 mm) Oregon timber rail was commonly used to provide overhead support for the equipment. Peter Freeman notes another exception to locally grown timber: 'A contemporary account of Cunningham Plains woolshed in 1872 mentioned that the entire floor was of "American" timber.'

The late 1800s were the 'glory days' of shearing and woolsheds. It was not unusual to have 200 000 sheep shorn in a single, huge shed in one season. The two sheds supposed to have been the largest in the country were Tinnenburra in Queensland and Burrawong in New South Wales, each of 101 shearing stands. It is claimed that Bowen Downs in Queensland shore 364 742 sheep in 1899, making it the world's largest sheep station, a position which it held for some considerable time. An even more impressive figure was divided between two sheds, both owned by the same man. In 1894, Dunlop station put through 284 000 sheep and its neighbour, Toorale, shore 220 000, yielding a grand total of over half a million sheep shorn in a single season. The shed at Brookong station, New South Wales, normally shore 7000 to 10 000 sheep per day.

Vagaries of the marketplace, wide fluctuations in the price of wool, weather conditions ranging from flood to drought, and governmental bureaucratic policies all combined to kill off the large sheep holdings. Where 100 000 population sheep properties once stood, 10 000 population stations now take their place. The day of the huge shed is past. The time of modest woolsheds seems set to last into the foreseeable future.

The design of a modern woolshed is a technical subject worthy of individual study. It has been estimated that there are in excess of 65 000 woolsheds presently in use in Australia. The number comes from the nation's 70 000 commercial sheep properties, 95 per cent of which are considered to possess a shearing shed. The role of a woolshed is simplicity itself. Sheep arrive at a set location, they are penned, shorn and dispatched. Their wool is then classed, baled and temporarily stored for shipment. That's it. Nothing more nor less.

TWO COMELY LADIES ADORN THE RANKS OF WOOLSHED WORKERS. NO LONGER UNCOMMON, FEMALE SHED HANDS ONLY APPEARED ON THE BOARDS IN AUSTRALIA IN 1983.

Most sheds follow one of two arrangements: an across-the-board layout or a centre-board layout. In the former, the shearer procures a sheep from a catching pen, dragging it across the board to be shorn. When fleeced, the animal is simply sent on its way to a counting-out pen. In the centre-board shed, the shearing board occupies a relatively central area in the shed where there is less movement required for shearers and shedhands.

Interestingly, when a sheep is on the shearing board it's the only time during its stay at the shed when it is on solid, unbroken flooring. The rest of the time the sheep is penned on slatted, battened floors which allow droppings to fall through to the ground below.

That no single layout is ideal for all situations is easy to illustrate. Authors Barber and Freeman discuss 'the display of shorn sheep', one of the factors in shed design:

> In the chute system, the shorn sheep move via chutes to under-floor races which lead them to counting-out pens where they are not readily visible to other shearers or personnel working in the shed. In the return race system, the shorn sheep are fully visible to others in the shearing shed. The significant elements in this matter are the shearer's pride of workmanship and his competitive spirit. Return races are held by some to encourage a better standard of work, whereas in a chute shed a shearer could be inclined to increase speed at the expense of quality.

Not even hinted at so far are the competing interests which must be accommodated for a shed to be efficient at any given property. Movement of sheep, dynamics of labour usage, economics of the building itself, requisites of employees, and hygiene needs of the sheep are only a few of the elements to be considered.

Sheep behavioural studies have shown that thorough, high-level lighting of an even nature facilitates the movement of sheep in and around the shed. Sheep do not like to move through darkened areas or those illuminated by contrasting levels of light. They do not like to see their shadow cast in front of themselves. Sheep move better around blind corners; they would rather move up an incline; animals made to look downwards are more secure; if sheep moving forward can see other sheep behind them they will attempt to stop and go back; a narrow race encourages sheep to fill faster than a wide one; separated sheep attempt to rejoin the main flock 'irrespective of the position of either man or dog'.

So with all these plus many more competing and conflicting factors to consider in designing a woolshed for the 1990s, it's much more enjoyable to find a historic, timber shearing shed and enjoy its milieu as well as its interior. Taken on a peaceful, sunny, non-working day, the medicine is a wonderful soporific.

AN ABANDONED SHEARING SHED FROM THE DAYS OF THE SOLDIER-SETTLERS ATTESTS TO THE EFFECTS OF DROUGHTS AND LOW WOOL PRICES.

WELL-WORN WOOLSHED AT CURRAWILLINGHI IN QUEENSLAND. SHEEP REACH THE SHED VIA A SWINGING BRIDGE ACROSS TRANQUIL BALLANDOOL CREEK.

10

Chef de BUSSHE *or* CROOK *Cook*

Two hundred years and more ago, we find the famous Spanish Merinos mustered for the shearing season. It is a gala event of immense proportions: 40 000 workers attend to 5 000 000 sheep which have been concentrated into a few huge shearing sheds.

The task took nearly a month to complete, assuming a celebratory nature among the hard-working and joyful participants. A worker's daily food allowance comprised one-tenth of a sheep (the system was one sheep per day for every ten men) and two pounds (0.9 kg) of bread. As well, wine stewards strolled among the toilers, distributing liquid refreshment to one and all. Since each person was 'allowed 18 draughts per day', according to contemporary account, it becomes immediately obvious why the shearing season was such a festive occasion.

EACH DAY A SHEEP IS BUTCHERED IN THE LATE AFTERNOON. TOMORROW'S MENU FEATURES *FRESH* MEAT.

There are certain similarities in the mess habits of early Spaniards and present-day Australians engaged in shearing. Ivan's crew consumes one freshly butchered sheep per day if the team numbers ten or less; over ten and a second animal is slaughtered. While there is no local equivalent to the Spanish consumption of almost a kilogram of bread per day it is recognised that Australian shearers eat tremendous quantities of food. And why wouldn't they, considering the utterly physical nature of the job?

The eighteen draughts of wine allowed each Spanish worker would last for a while in most local shearers' hands but that same assurance wouldn't hold for stubbies or tinnies. In the refuse barrels placed outside the shearers' huts, bottles and cans are a major, ever-expanding part of their leavings.

The search for good food and a good cook to prepare it has always been an important goal of a shearing team. Sometimes that goal has been elusive.

The oldest story in the bush deals with this situation. It seems that a shearing team was gathered together discussing the many deficiencies of their present cook. The boss came around, joining in the lively debate. He put a question: 'What I want to know is who called the cook a bastard?' From among the men came an immediate rejoinder: 'What I want to know is who called the bastard a cook?'

Both the quality and quantity of tucker are important factors when the strenuous working day begins at 7.30 a.m. and ends ten hours later. The shearing members of a team point out that they are entitled to good food since they must pay for their meals. It's true. The Federal Pastoral Industry Award sets separate wages for 'found' and 'not found' workers.

Shearers living away from home, as those who work in Ivan Letchford's team do, are found. An amount of nearly $80 is deducted from their wages each week to pay for their meals on a seven-day basis. Whenever possible, everyone goes home for the weekends and this means that cooking for Ivan's team is most commonly a five-day job.

Early pastoral workers in Australia were often shepherds. Their lonely lives reflected the importance of their rations for they were often nicknamed the 'Ten, Ten, Two and a Quarter Men'. This meant that each week they received 10 pounds (4.5 kg) of meat, 10 pounds of flour, 2 pounds (0.9 kg) of sugar and 1/4 pound (0.1 kg) of tea. Patsy Adam-Smith quotes an observer of the times: 'They lived on damper and mutton year in, year out. The nearest approach they had to vegetables was pigface, horehound and marshmallow that grew around sheep camps. They reckoned any edible vegetation tended to keep off scurvy, of which they were always in dread.'

Skipping ahead a century and a half to the current award applying to the pastoral industry, we find that the Australian Conciliation and Arbitration Commission lists the following Scale of Rations for found workers:

Bread or flour
Meat
Vegetables, potatoes, onions, beans, peas (split and blue), green vegetables (when reasonably procurable)
Oatmeal
Rice
Cornflour
Tapioca or sago
Macaroni
Barley
Jam
Fruit—currants or raisins, dried apples, apricots,
Prunes
Sugar
Syrup
Honey
Milk fresh, condensed or powdered
Spices, herbs, pepper
Essence
Pickles
Vinegar
Sauce
Soap and washing soda (for cleansing cooking utensils)
Bi-carbonate of soda
Cheese
Butter (when reasonably procurable)
Suet
Cream of tartar
Eggs
Dripping
Jelly crystals
Custard powder
Tinned fruits
Fish
Tea, coffee or cocoa
Curry
Salt (fine)
Mustard

One doesn't need a degree in dietetics to notice that this diet is quite deficient in fresh fruit and vegetables. Extreme sustained heat and remote locations explain the notation 'when reasonably procurable' after 'green vegetables' in the Scale of Rations. Since modern nutritional science tends to advocate large quantities of complex carbohydrates in the diet, the rations for shearing industry workers fit right in with this aspect of current thinking.

In the 'good old days' of the 1800s, cooks were close enough in duties to general hands. One chap ran into a farmer who was looking for a cook. C. E. W. Bean continues:

> The manager asked what sort of a cook he would prefer. 'Oh, well, just a man who can kill a sheep and cut wood,' he said, 'an' do a bit of milkin', and perhaps fetch water. It might be a good idea if he was a good hand with a horse ...'

Still in the nineteenth century but somewhat later than the shepherds' time, we come to the era of the swagmen. These men—there were few female swaggies—wandered over the face of Australia well into the 1900s. Two main types of swagmen populated the bush: the itinerant workers who travelled over the countryside in search of work, sometimes covering a territory in the manner of a circuit rider, were simply a mobile labour source. Quite frequently they were shearers and shedhands. The second variety, many of whom came to be known as 'sundowners', tramped the byways in search of an easy life in much the same way as today's dole-bludging drop-outs shun any kind of regular work. In either case, these people had to eat. Edel Wignell writes:

> The staple diet in the Australian colonies in the early days was mutton, damper and tea. Everyone, including the swagmen, soon grew tired of it. Some people began to eat 'bush tucker' as well, fish and game, snakes and goannas. From the Aborigines, the bushman learnt how to use the hot ashes of the camp fire as an oven for cooking both damper and such game as pigeon and wallaby ... Bush foods such as snake, possum, goanna, dugong, witchetty-grubs and turtle eggs were baked in a similar fashion. Yabbies, caught in a waterhole or creek, were a tasty treat.

Wignell also discusses the early damper which comprised flour, water, salt and—when available—baking powder. The ingredients were 'mixed to a firm dough, which was shaped into a flat cake and dusted with flour'; after baking in the ashes and/or coals of a fire, the result was a variety of everyday bread. Brothers Jack and Reg Absalom are authorities on the cast-iron or pressed-steel camp ovens, described by them as 'the most used cooking utensil in the outback of Australia since the first settlement'.

The Banks of the Condamine

Man:
O hark the dogs are barking, love, I can no longer stay;
The men are all gone mustering, and it is nearly day.
And I must be off by morning light, before the sun does shine,
To meet the Sydney shearers on the banks of the Condamine.

Girl:
O Willy, dearest Willy, O let me go with you!
I'll cut off all my auburn fringe, and be a shearer too;
I'll cook and count your tally, love, while ringer-o you shine,
And I'll wash your greasy moleskins on the banks of the Condamine.

Man:
O Nancy, dearest Nancy, with me you cannot go;
The squatters have given orders, love, no woman should do so
And your delicate constitution is not equal unto mine,
To withstand the constant tigering on the banks of the Condamine.

Girl:
O Willy, dearest Willy, then stay at home with me;
We'll take up a selection, and a farmer's wife I'll be.
I'll help you husk the corn, love, and cook your meals so fine
You'll forget the ram-stag mutton on the banks of the Condamine.

Man:
O Nancy, dearest Nancy, pray do not hold me back!
Down there the boys are mustering, and I must be on the track.
So here's a goodbye kiss, love: back home I will incline
When we've shore the last of the jumbucks on the banks of the Condamine.

Their damper is made from six cups of self-raising flour, a tablespoon of baking powder, a pinch of salt and warm milk. The mixed ingredients are placed in the camp oven and surrounded by hot ashes.

It is now time to introduce Debbie Lumsden, who decided to sign on and see the world as a cook for Ivan's shearing team. During Debbie's tenure with Ivan I wrote 'The Shearers' Cook', which originally appeared in *Family Circle* magazine. *Family Circle* is hereby thanked for permission to reprint this abridged version.

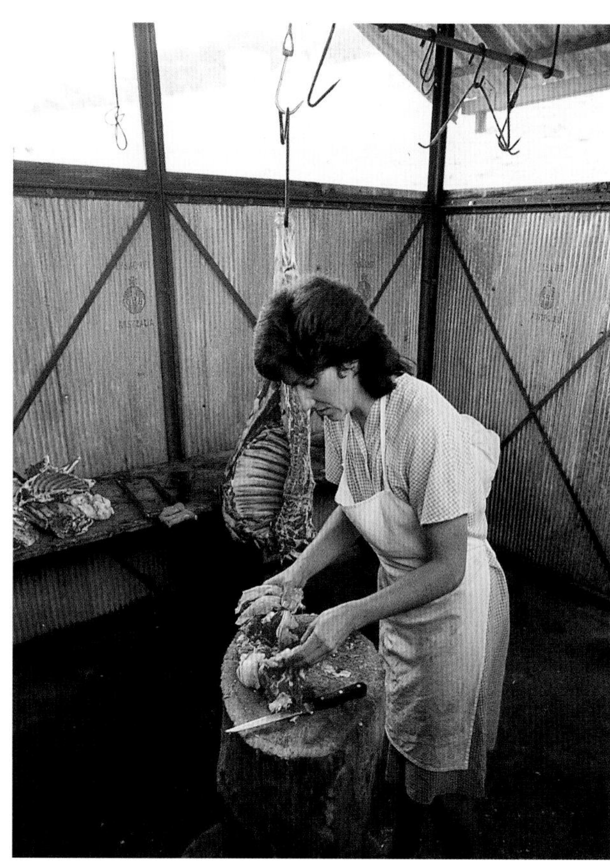

JUDY GARDENER AND DEBBIE LUMSDEN BREAK DOWN SHEEP CARCASSES IN MEAT HOUSES.

LYNN TOLMIE CAN COOK FOR A TEAM OF UP TO TWENTY-FIVE PEOPLE ON THIS FUEL STOVE. THE FIREBOX CAN TAKE THREE-FOOT LOGS, SIX INCHES IN DIAMETER.

GIDGEE WOOD FOR THE COOKS' STOVE CUT FROM ONE-HUNDRED-YEAR-OLD FENCE POSTS

The Shearers' Cook

'I was on my way to a job as a station cook. When I met the local publican and asked him where the property was located, he began to laugh. He said they'd had six cooks in as many months and the owner's wife was impossible to work for. The publican was right. After five weeks I quit and had to find another job. There was an ad in the paper for a shearers' cook and so I answered it. The job was mine and I've been a shearers' cook for three years now.'

Debbie Lumsden is telling me of her background while she is baking pans of biscuits. She is twenty-eight years old, slender, pretty and feminine, not exactly what one expects to find in the midst of a group of rough-and-tumble shearers.

Ivan Letchford is a traditional shearing contractor who has been going to some New South Wales properties for more than thirty years. Ivan is one of the last contractors to still employ a permanent cook; his team must be fed five times a day, rain or shine. Debbie Lumsden does the job single-handedly. The smallest team she has cooked for contained twelve men, the largest, thirty.

Debbie's job generally begins at 4.30 a.m. and ends about 7.30 p.m.; fifteen hours of effort which makes the hard-working shearers' day seem relaxed in comparison. The shearing shed works four two-hour blocks per day: 7.30-9.30 a.m.; 10 a.m. until noon; then 1-3 p.m and 3.30-5.30 p.m. Debbie's job requires that she fit into this schedule preparing three hot meals and two cold feeds for the shearers' breaks.

Here is an outline of a typical day for Debbie:

4.30 a.m. Arise, light wood cooking stove, get it hot so breakfast can be served at 6.30 a.m. By breakfast time have breakfast cooked, vegetables for lunch prepared, and sheep 'cut down' (reduced into appropriate cuts—more on this later).

6.40 a.m. Sit down to breakfast and a cup of tea.

7.00 a.m. Wash up from breakfast; do baking, clean out meat house.

9.00 a.m. Stop whatever doing to make sandwiches and morning tea for men.

9.30 a.m. Sandwiches, cakes, biscuits and tea to shearing shed for men. Have break with them.

10.00 a.m. Make dessert for noon meal if it is to be a hot one, or for next day if it will be a cold dessert.

Lunch cooked.

Start preparing soup and portions of evening meal. Advance menu planning.

12 noon	Serve lunch to shearing crew; then have own meal.
12.30-1.00 p.m.	Wash up lunch dishes, then prepare food for afternoon tea.
3.00 p.m.	Back to shearing shed with tucker for afternoon tea. Have break with men.
3.30 p.m.	Complete preparation of evening meal.
5.00 p.m.	Drink time! Nightly glass of wine.
5.30 p.m.	Chat with boys at end of shearing day.
6.30 p.m.	Serve dinner; then have own meal.
7.00 p.m.	Wash up/clean up.
7.30-8.00 p.m.	Finish. Free time until bed.

This schedule is followed five days a week if the crew is close enough to return to their homes for the weekend. If they are out in the backblocks, too far away for a weekend trip home, the men have Saturday and Sunday off, as per union rules. Nevertheless, Debbie must prepare three meals per day for them.

The shearing contractor, Ivan, remarks: 'If I could just get Debbie to work during the lunch hour, I'd be able to obtain twenty-five hours per day out of her.'

Debbie prepares her menus one to three weeks in advance, depending on the job site, and Ivan does the food shopping on the weekend. Some of the staples required for a crew of twenty-five to thirty men for a five day week include a 60 kilogram bag of potatoes, 24 loaves of bread, 16 kilograms of margarine, 12 kilograms of sugar, 10 kilograms of plain flour, 12 kilograms of self-raising flour, 12 kilograms of tomatoes, 4 pumpkins and 10 sheep.

The property owner supplies the meat for the shearing team. Since the most readily available and economical source of meat is sheep, invariably the meal features mutton. Depending on the size of the shearing team, one or two sheep are butchered every evening. After the sheep are slaughtered they are hung overnight in the meat room where Debbie sections them into usable cuts the following morning. Because mutton is the only meat served, Debbie has to be a magician in preparing it in a variety of tasteful ways. They have mutton steak, mutton chops, baked mutton, mutton mince, curried mutton, marinated mutton, mutton meat loaf, mutton rissoles, 'corned' mutton and mutton sausages. She even has a recipe to make mutton taste like ham and another for mutton in spaghetti bolognaise. It's definitely not true that some shearers have begun to bleat and grow a woolly coat!

Everyone who has enjoyed Debbie's cooking can vouch for the fact that she presents meals which are delicious and hearty. A sample breakfast comprises savoury rice, creamed corn, steamed tomatoes, toast and coffee or tea. Other breakfast items include eggs, rissoles and sausages. Lunch could be cottage pie, mashed pumpkin, braised cabbage, ice-cream and jellied fruit. Dinner could be beef and tomato soup (read 'mutton' for 'beef'), marinated mutton steak, sauteed potato and tossed salad. Aside from the ice-cream, desserts include apple pie, steamed pudding, self-saucing pudding, flummery or apple snow. Cakes and biscuits supplement the sandwiches at morning and afternoon breaks.

Debbie reminds me that, 'You must do everything yourself—there is no corner store for a shearing shed!' The mince isn't purchased from a shop, it's made from a freshly slaughtered sheep. The creamed corn is made in a blender, the ice-cream homemade and then chilled in their own freezer, and the sausages fashioned in their own sausage maker. Ivan bought a freezer which they transport from job to job on the back of his truck, enabling Debbie to provide more variety in her meals.

What appliances does Debbie have to assist her in preparing her flavourful meals? 'In the course of a year's work two sheds have gas stoves; all the others have wood stoves. The only electrical appliances are fridges, where there is electricity. Some places don't even have electricity and kerosene fridges are used.'

Kerosene fridges are a particular irritation to Debbie: 'I hate them. Thirty years ago they were a wonderful invention but not any more. The parts are worn or broken and they can't be replaced because the fridges are no longer manufactured. They're not efficient, especially when you are working in 30°C temperature—either they freeze everything or the butter runs out at your feet.'

Hot water for showers again comes from a wood fire. Water is heated in a 'donkey boiler'—a couple of 44 gallon drums stuck together with a fire beneath it to heat the water.

Firewood for the stoves is supplied by the

property, cut to 'stove-sized length', according to the usual contract. Debbie says: 'It doesn't matter how round it is, just as long as it's not too long to fit in the wood box. Usually I have to split it myself and then cut the kindling. So in the afternoon, when I have a little time, I get out with an axe, split some wood and cut a bit of kindling.'

Debbie sent an article in to ABC radio which was aired on the 'Australia All Over' program. It was called, 'Why Cooks Are Cranky Old Bastards' and it was about 'the rotten old bloody fuel stoves you have to work with'. Some stoves are 'cold' and cook poorly while others are 'hot', not only burning the food but burning out the cooking tray as well. Other stoves are tilted so that cakes come out looking like the Leaning Tower of Pisa. 'One stove I work on,' Debbie relates with a wry laugh, 'is just impossible. If you make a casserole the gravy runs out the back of the dish to one corner of the stove, down the edge, and comes out at your feet.' Moreover, some stoves must be built up with bits of brick to level them, or you have to use foil to contain the food juices.'

Her only electric appliances are her own, comprising a blender, a Mixmaster and an electric mincer. She packs them up and takes them along from job to job.

Debbie has been a cook all her working life. Growing up in Blacktown when it was more of a country town than a Sydney suburb, she joined the Navy after high school. She spent five years as a Navy cook, then had jobs as a station cook, did pub cooking near Lithgow, New South Wales, and was a camp cook at Fraser Island, Queensland. The latter involved cooking for people being introduced to safari or tourist camping via five-day stays on Fraser Island in the tourist season.

During her three years with Ivan Letchford and his team, the comely, hazel-eyed brunette has plied her skills in the New England, Hunter and north-west regions of New South Wales.

As with all occupations, cooking has its pluses and minuses. Debbie likes her work and speaks freely of the pros and cons of being the only woman amongst a group of rough-hewn males. 'Cooking is the easy part. You must be a child psychologist to deal with them successfully. It's hard being a shearers' cook because you don't just work with them, you live with them.

'In a normal job, after eight hours you knock off and go home to your family. Here you see them at night, you see them in the morning, you see them at lunch. You see them twenty-four hours a day and after a while you get sick of each other. Their little idiosyncrasies begin to bug you and yours begin to bug them.'

On all her free weekends Debbie returns to Sydney to spend time with family and friends. 'My biggest treat is to be able to sleep in until 6.30 a.m.'

Asked why she chose to take such a demanding job, Debbie sums up her feelings in one sentence: 'I like the bush atmosphere; I like country life and country people '

The shearing season ends before Christmas and the entire team enjoys a break of approximately six weeks' duration over the summer holiday period. While she is on the job, Debbie enjoys reading, sewing and bird-watching in her leisure time. 'If there's a river on or near the property we are working on, I like to go swimming.'

Considering the hours worked, Debbie believes the pay is poor but there are no distractions to tempt you into foolish shopping trips. 'If you want to save money for a specific reason, it is a good job to do for about three years. Because your time is totally tied up

FOR SOME UNKNOWN REASON, IN EACH MESS EVERYONE MAINTAINS THE SAME SEATING POSITIONS. IF ONE WISHES TO CHANGE SEATS, IT IS CUSTOMARY TO WAIT UNTIL THE START OF THE NEXT SHED.

AN UNSEEN SHEDHAND HAS JUST THROWN HIS FLEECE ONTO THE CLASSING TABLE FOR SKIRTING.

in work, there is little opportunity to spend any money. Besides, you're completely found–you have a bed to sleep in and you're well fed–so your bills are virtually nonexistent.'

When Debbie was asked what qualities she would look for if she was to hire a shearers' cook she listed four characteristics as absolutely necessary for the job:

1. You must be terribly easy to get on with.
2. Nothing whatsoever could upset you.
3. You must be able to get along with all kinds of weird and wonderful people (more weird than wonderful!).
4. Hard work and long hours must be acceptable.

The sheep-shearing industry in Australia is an anachronism in modern society, according to Debbie. She observes, 'When Captain Phillip arrived in Sydney, he unloaded a couple of old wood stoves and a few shearers and they started in shearing. There's been almost no change in the industry in the last 200 years. We still have the same equipment, the same appliances, the same food and even, I believe, the same stories as they had last century.'

Discussing food led Debbie back to the present and the multiple jobs she was doing as we talked. What about the leftovers? In today's country living there is no room for waste. 'If mashed potato is left over I can make a 'Pioneer Boston bun' out of that. The mashed potato is added to sugar and flour, no yeast is used and the result resembles a teacake, bread-like in flavour and texture. Or, I can add a bit of onion, egg and flour and make "Irish bread".'

Debbie points to the pumpkin she is cutting up. 'I use the leftovers to make pumpkin fruit cake or pumpkin soup. With vegetables, generally I mix them with eggs, creating a variety of scrambled vegetables, called 'bubble-and-squeak'. Cold meat is made into sandwiches or minced, yielding cold meat croquettes. Casseroles and stews have not ended their useful life. Stew can be placed in the blender and with a bit of curry powder added in, the result is mulligatawny soup!'

Questioned about a favourite recipe, Debbie immediately mentions 'quick biscuits'. 'I like them because they are easy; the shearers like them because they are big.' The recipe yields five dozen biscuits and feeds a team of twelve men twice in one day, or twenty-four shearers at one meal.

The recipe for 'quick biscuits':

Melt 375 g (12 oz) butter
Add 425 g (14 oz) sugar
Add 625 g (20 oz) self-raising flour
Add 2 eggs
Add your fancy for flavouring, for example, sultanas, cocoa, or coconut
Drop dessertspoonfuls of the stiff dough onto an ungreased tray.
Bake at 400°F for 10 minutes.

As this article is being completed Debbie discloses that she is finishing up as a shearers' cook at the end of the current season. She plans to return home where she will embark on a three-year course in fashion design. Her goal is to establish a business of her own, 'designing clothes for the individual'. How will she support herself during the long three-year course?

'I'll probably be a cook!'

SHEEP WAITING TO BE COUNTED OUT. SEVERE WEATHER CONDITIONS LEADING TO MATTED FLEECES HAVE MADE LIFE MORE DIFFICULT FOR SHEEP AND SHEARERS.

Tucker Tales

IVAN
Old Jim was a tough one all right. This mad cook we had owned a pet magpie and a pet wallaroo. When I went home this one weekend he went right off his rocker. When Jim and I got back on Sunday night, there was a bloke waiting for me at the gate and another couple of guys camping out in the shed. This cook had gone berserk and he had all the men off hiding in the hills. He was armed with a meat cleaver and everyone was afraid of him. Now Jim was a tough nut and he said, 'Where is this bastard?' The cook was in his room shouting, 'Don't you come near me! Stay away from me!' All the while he's brandishing a cleaver and an axe. Old Jim just walked in and said, 'Gimme that axe and gimme that cleaver you bastard or I'll give you a kick up the arse.' That was the end of it. The rest of the blokes had spent the entire weekend up in the hills. Oh yes, Jim was a tough one.

A SHEARER
Remember the cook who used to put tinned fruit on for tea and there was never any juice with the fruit? Well, he was pissed all the time and he was putting the juice from the tinned fruit into his metho.

A SHEARER
The cook said to Ernie, 'Did you get the seven-pound tin of Sao biscuits?' 'No,' Ernie said, 'I couldn't. On the side of the tin it said, *'Are not Sao biscuits.'*

IVAN
Someone went over there and Ernie had biscuit tins piled up all around the place. There were dozens of Monte Carlo biscuit tins piled everywhere.

'You must really like Monte Carlo biscuits, Ernie.'

Ernie says, 'They're good dog tucker.'

'Don't tell me you feed your dogs on Monte Carlos?'

'Well,' says Ernie, 'You tell me where I can get cheaper dog tucker than 2 bob a tin.'

'That's not what they cost, that's the deposit on the tins you silly bastard.'

IVAN
One day at Midgee in the Pilliga Scrub country in the days when we were really concerned about money, I went to the toilet. Looking down, I saw that the cook had thrown a couple of plum duffs down there. He must have ruined them and in that black country there's no gullies to hide them in. The men decided to sack this cook and said that we'd get rid of him on Friday evening. He was a big, crazy bastard that everyone was afraid of because he had already threatened to shoot some blokes. I wasn't game to sack him so Allan says, 'We'll borrow that ringer's horse and ride over together.'

We got over there and Allan yelled out, 'Are you there, Cook?'

He answered, 'Yes.'

Allan hollered back, 'You're fired.' We turned and cantered safely back to the shed.

DEBBIE
When asked about priority rules for shower times for cook and shearers:
I think it was first in, best dressed. If I had been in the shower when they knocked off work I suspect they'd all have walked in and taken a shower as well, without even noticing I was there.

IVAN
Years ago in the big sheds, the shearers had a different mess to the shedhands. The shearers' cook belonged to the shearers and the shedhands had their own cook. Each one had a different smoko box and problems arose at times because one or the other had a better cook: Today's cooks have it easy compared to the old days. There was no such thing as a fuel stove or a range until the 1940s. Before that we had brick ovens. You'd put six-foot (about 1.8 metre) logs in the oven—it was like a brick kiln—and then you'd withdraw the coals in the morning. Outside, you'd cook the chops you'd removed from the brick oven and you also cooked your vegetables—on a galley. The cook would put the meat in the brick oven and as it cooled down he'd do his meat pies, cakes, apple pies, etc.

A Shearer

Who was that big, dumb blond-headed kid who wasn't much of a rouseabout? We got him to do the washing-up every night. About the only thing he could do properly was to clean up the pots and pans. We sent him down to make the tea and all the leaves were just floating on top of the water. Someone said, didn't you wait for the water to boil? 'No,' he said, 'it says on the packet put in pot and add water.'

A Shearer

Some of my most pleasant memories are of weekends in Queensland in the wintertime, on the sunny side of the huts. The cook spoiled you on the weekends after five days of hard shearing. It was part of the recuperation process ... the feeding, resting and reading. I wouldn't even go fishing just in case I caught one.

Patsy Adam-Smith in 'The Shearers'

Jim Carem heard many things said to and about cooks. 'One of the shearers I know was caught in Singapore by the Japs and thrown into a POW camp. The Japs gave them muck to eat. Men were dying of it. A bloke says to this shearer, 'You don't seem to mind this muck?' 'Why should I?' the shearer said. 'I've lived in shearing sheds half my life.'

I was in a shed once that had a real dirty cook. One day the shearers bawled out, 'Cook! There's maggots on this meat!' The cook bawled back, 'There ain't! I swear I wiped them all off.'

Dennis Ryan

I remember a shed at Cunnamulla with fifteen shearers and thirty-five on the team all up. There was a mix-up with the cooks and we had to put up with a bloke who was bloody horrible. So when we came to the next shed in Charleville, we were all just interested in the cook. I said to the boss, 'I can't put up with another shed and no tucker.' He said, 'Don't worry, we'll have a good cook for you.'

We made enquiries and scouted around. We found out there was this bloke called Singin' Jim Slattery, who was a cook, but that we'd have to work on him. We were to get him pissed and get him started talking ... ask him what can he do? How does he do this? How does he do that? We had to get him to talk his bloody head off and then tomorrow morning when he wakes up at the shed with a hangover, we'd remind him of everything he'd claimed he could prepare.

Sure enough, it worked beautifully. We killed a beast and had steak every night; we had Spanish cakes and there was cocoa for supper as well as cold sweetbread, evenings. He had a specialty of homemade mutton sausages and he baked fresh bread daily. Oh, it was wonderful.

Tim's problem was that he was an alcoholic, which most of them were. A lot of them didn't drink in the shed, they only drank between sheds. They used to make a lot of money if they were good cooks. They'd get today's equivalent of several thousand dollars for a shed. They wouldn't drink for perhaps six months when they were working. Then after a drunken bout, they'd ring someone and say, 'I'm right for a shed now'. They were dependable, too.

Many of them were also gamblers. Now if you had a bloke in the shed who was a gambler and he'd won at the weekend, the cooking was sensational. If he'd lost though, by Christ we'd nearly starve to death. You see, he was suffering from remorse as well. There really was a lot of psychology in being fed.

In this day and age a cook similar to Singin' Jim Slattery would be cooking at an RSL club. The other thing was, as these blokes were alcoholics, the bush was like a desert island to them. They were secure there, well protected, they made good money. They could cover up and be their alcoholic selves without responsibility. Then, when they needed responsibility again, they could settle down and away they'd go.

Debbie

The grazier's wife was a witch. She was a real whinger, nothing was ever right. It was like working for Hitler. Why, employees even had to be in bed by 10 o'clock nights. She had her husband under her thumb unbelievably—the men called him 'pussy whipped'. He was in a bad way.

I was cooking for eight people. She would write up a menu for me to follow. I never followed it terribly closely, it depended on what food was left over, or what vegetables there was a glut of—I'd make my own decisions.

The crunch came one night when she had two

desserts written down. Oh, this is stupid, I thought, making two desserts for eight people. So I made only one. She went off her brain at me in front of the men and that was it. I said, 'You can take your job and stick it, I don't need it.' I was told I'd been one of their longer-lasting cooks.

A Shearer

Each week, two or three of the boys said the curry wasn't hot enough so this time the cook made two curries, one for them and one for us. Their curry was so hot it was green and they had to sip water between bites. They wouldn't give in that it was hot enough so the cook left the rest in the pot, planning on giving it to them for breakfast. In the end he couldn't because he left the spoon in the pot and overnight it took all the nickel off the spoon.

Ivan

Old Jim was cooking then and the men were getting pretty hungry. They came in and sat in the dining room and started up a rat-a-tat-tat against the side of the table with their knives and forks. 'Will it be much longer, cook?' Rat-a-tat-tat. 'How's she going, mate?' Rat-a-tat-tat. 'No,' Jim said, 'it won't be long now.'

When the meal was cooked just right, he pulled the lids off the pots and threw all the food in the middle of the dining room. 'There it is,' Jim said, 'I'll give you bastards rat-a-tat-tat!'

A Shearer

When I walked in I said to the cook, 'What's for tea?' She answered, 'Soup and stew.' I only saw the one pot there and so I said, Where's the soup?' She said, 'The stew is on top, the soup is underneath.' There were four chops in the top of the pot. When you took those out she reckoned that was the stew and the soup was the liquid beneath.

Debbie

I used to enjoy a drink as well, so I'd call 'soup' at half-past six and sit out on the veranda, sipping my wine. The nights would go on a lot later. Half the team would be in bed and asleep and there'd still be a group of us sitting up at 10 or 11 o'clock at night. If you don't have to get up and go off someplace to work, I can't see the point of going to bed at 7 o'clock at night. Especially when all you have to do is get out of bed and do your job. You may as well sit up and enjoy the bush, the smells and the quiet.

11

A DRINK *or* TWO

There has been an association between shearers and grog going back for centuries. Mentioned earlier was the Spanish habit of employing wine stewards who went around serving drinks to one and all at shearing time. In colonial days, shearers received alcohol as part of their wage structure. There was a time during the gold rush in Victoria when it was nigh on impossible to get competent shearers. Those who were procured charged the exorbitant rate of 35 shillings per hundred plus a premium of two glasses of grog a day.

Turning north to Queensland, author Judy Harrison discovered a quaint article in the *Longreach Herald* from November 1876. The editorial opinion comments on the undesirable lifestyle maintained by shearers and ties their objectionable ways to the evils of strong drink:

> If shearers were a thrifty class and saved their money or turned it to some good account, we might be more inclined than we are to back them for standing out for the old rate. But when we remember that three-fourths of the money goes straight to the public houses and does the poor shearers more harm than good, we cannot regard it as anything but a public benefit that their wages should be reduced.

'FINISHING OFF FINAL BLOWS ON LAST HIND LEG'. SHEARING OF THIS SHEEP IS ALMOST COMPLETED.

The last two sentences of the same article undoubtedly would have brought forth hearty guffaws from those Antipodean shearers: 'The saddest feature about them is perhaps their total lack of domestic ties. They seem to be a nomadic race who wander and wander and never know the charms or soothing influences of "Home Sweet Home".'

Obviously, the matter of alcohol and shearers engenders differing viewpoints. It all depends on who is rendering the social commentary. Our old friend Duke Tritton sees it in a different light from those pristine wowsers at the *Longreach Herald*:

Bluey Brink

There once was a shearer by name Bluey Brink,
A devil for work and a devil for drink;
He could shear his hundred a day without fear,
And drink without winking, four gallons of beer.

Now Jimmy the barman who served out the drink,
He hated the sight of this mad Bluey Brink,
Who stayed much too late, and came much too soon,
At evening, at morning, at night and at noon.

One morning as Jimmy was cleaning the bar
With sulphuric acid he kept in a jar,
Old Bluey came yowling and bawling with thirst,
'Whatever you've got Jim, just hand me the first!'

Now this ain't in histories, but men swear it's true,
Bluey drank acid without cracking a boo.
Saying, 'That's the stuff, Jimmy! Strike me stone dead,
This'll make me the ringer of Stevenson's shed!'

Now all that long day as he served out the beer,
Poor Jimmy was sick with his trouble and fear;
Too worried to argue, too anxious to fight,
He pictured poor Bluey a corpse by the night.

But early next morning, he opened the door,
And along came old Bluey, asking for more,
With his eyebrows all singed and his whiskers deranged,
And holes in his hide like a dog with the mange.

Says Jimmy: 'And how did you find that new stuff?'
Says Bluey: 'It's fine, but I ain't had enough;
It gives me great courage to shear and to fight,
But why does it set all my whiskers alight?

'I thought I knew drink but I must have been wrong,
For what you just gave me was fierce and strong,
It made me cough, and you know I'm no liar,
Each bloody cough set me whiskers on fire.'

Having been a shearer, I may be prejudiced, but I think the shearer is justified to a certain extent in boasting of his prowess. Being an individual worker, his earnings depend entirely on his shearing ability. Handling such an unpredictable animal as a sheep, quiet one moment, kicking and struggling the next, he has to think and act quickly to avoid cutting it or knocking it about. And he must never lose his temper, even if the boss of the board is on his back all day. So the shearer may be forgiven if he relaxes over a few grogs.

Relaxing over a few grogs is one thing. Being paralytic and comatose is another. Enter the subject of lambing down. During the 1800s, owners would send station hands out into the paddocks when the ewes were preparing to give birth. This process of checking on the welfare of the sheep was called 'lambing down'. Through an unknown transfer of idiom the phrase came to refer to the way many shearers spent their pay-packets. Going to a hotel, turning the entire pay-packet over to the publican and remaining there, drinking for as long as it took to exhaust the whole year's earnings, became known as lambing down.

It was common, it was senseless and it was long-lasting. C. E. W. Bean says that 'at the time of World War I, "lambing down" was becoming less common, but it had by no means vanished'. Patsy Adam-Smith provides a lucid, terse comment on the practice: 'In retrospect it appears a tragic waste of the effort of those men who carried out an immense amount of brave, monotonous, hard work year after year for nothing more than a few days' feverish drinking.'

The basic lambing down procedure is described above. Variations on this procedure included the old drugging-the-drinker scheme so that he thought he had consumed gallons of grog; actually he had received a Mickey Finn. Another variant found the revived drinker awaking, surrounded by empty champagne bottles. He had supposedly bought drinks for every able-bodied person in the shire. Hector Holthouse tells how women were involved in assisting the shearers to knock down their cheques. A woman who kept a shanty near Winton took a man's boots when he first went to sleep. The boots were returned only after his cheque was used up. Some of the lovely *femmes fatales* of the day included Hairy Mary of the

Condamine; Black Alice, who was adorned 'with teeth like a Moreton Bay shark'; and Cunnamulla Mary, believed by Holthouse to be poet Barcroft Boake's 'woman in scarlet and black' because she was generally attired in a black skirt and scarlet blouse.

As the saying goes, 'that was then and this is now'. Things have changed markedly and shearers do not work for an entire year to give the proceeds to a publican in a hotel or a harpy in some outback drinking shanty. Still, they do enjoy an occasional drop and there is a philosophical overlay to this phenomenon, at least according to Dennis Ryan. He says that 'shearers just have to drink; if they don't drink they can't work'. Challenged on this point, Dennis responds:

> I can tell you about one chap to illustrate the point. He liked to drink and to fight. He was a decent shearer on Monday and Tuesday and that was about it. One time when he owed Ivan money he borrowed someone's car, got picked up for drink-driving and spent the night in gaol. Once he got out on bail he shot through. Eventually, after several of these episodes—he was like the Don Giovanni of drunken driving—he ended up in West Australia
>
> After a while we got a letter from him written on Her Majesty's stationery. I think it was a prison farm or something like that. The chap came from a well-placed family and this letter was quite well written. His purpose in writing was to say that he had a shed lined up for his release and could we have the necessary gear waiting for him then.
>
> At the bottom of his letter he said, 'Dennis, I want to tell you that I've really learned the error of my ways. I can see that in the past I've got into heaps of trouble from mixing driving and drinking. I can see now that it just doesn't work, especially for someone like me. When I sit here at night, lonely, I mull it over to myself and I have come to the conclusion that I'll be safe forever if I just follow my best intentions. From now on I'm never ever going to drive again.'

So it would seem that shearers are irretrievably linked to intemperance. This message came through clearly in an unknown author's anecdote that was jotted down. It concerns a time when shearers were hard to come by and getting a team together was not an easy task. The contractor arrived at one station and was asked, 'Have you got a full team?' He said, 'They are as full as I can get them', as he watched the drunks roll off the back of the truck. A suspicion arises that this introspective approach to shearers and drinking could be in error. Perhaps the best method is to read some imbibers' yarns solely for their entertainment value. For the rest of this chapter Ivan and the shearers regale us with tales from the dipsomaniacs' chronicles.

Drinking Stories

Ivan

We were shearing up about 4000 feet (about 1200 metres) and it was cold and snowy that winter. I lifted the bonnet on the old Holden and there's water in the oil. I said to Jim, 'I've done the head gasket, this won't get us to Armidale, let alone Walgett. I'll have to get another head gasket and get it fixed.'

'Wouldn't that just stuff you', Jim replied. 'I'll light the fire and you go and get the gasket.'

I went to Ebor and there's a pub there and I met a couple of blokes. They said, 'You'll never get a gasket in Ebor', so we got a gutful of booze into us and headed off for Grafton, eighty miles (about 130 km) down the coast. In Grafton we had a night on the booze, then we got the gasket, came back to Ebor and had the rest of the day there. I had been gone three days. I was too frightened to go back, so I had another drink and kept putting it off.

Finally, I took him a bottle of rum and a couple of bottles of beer and there he was standing in the doorway when I got there.

'Well,' he said, 'where have you been, you bastard?' He was a beaut old bloke. You know, he'd forgive you in a matter of minutes.

ALL THAT REMAINS OF A BUSH MOUNTED POLICE STATION ERECTED AFTER WORLD WAR I. SOLDIER-SETTLER BLOCKS OF 1280 ACRES (2 SQUARE MILES) WERE ORIGINALLY WELL-TIMBERED. AFTER THE NEW OWNER WENT ROUND WITH AN AXE, RINGBARKING TREES, BURNING TIMBER AND DIGGING UP SUCKERS WITH A MATTOCK, THE PROPERTY WOULD ONLY CARRY ABOUT 800 WETHERS, TOO FEW TO BE ECONOMICALLY VIABLE. THE POLICE PRESENCE HELPED TO PROTECT AGAINST SHEEP AND CATTLE DUFFING. IT WAS ALL TO NO AVAIL AS THE UNDERSIZED PROPERTIES WERE GRADUALLY SOLD OUT, CONSOLIDATED AND PASTORAL COMPANIES FORMED.

Angus had an annoying hernia and he was at the Victoria Hotel in Goondiwindi. There were always pubs in the bush, never motels. They were well known, they were your bankers, your guardian, they were everything; your mail might come there and you could leave your money there. Now, Angus says, 'This hernia is giving me bloody hell. I look up and there's two monsters (police) standing over me. I made a quick decision: I jumped straight up and banged both their heads together. I got three months in Boggo Road Gaol and had my hernia fixed up there.'

A Shearer

Is he retired now? Retired? He's dead, the poor bastard. Geez, he liked big, rough sheep. One night he went to the ball at Burren Junction and played up terribly. He came back to the hut in the early hours of the morning and got into his shearing gear. He had half an orange for breakfast, shore forty-four on his first run and then announced, 'forty-four on half a bloody orange'. With that, he put the handpiece down and went to the hut to sleep it off for the rest of the day.

Ivan

You used to start a shed only on mail days. Here there used to be mail three days a week. It was brought in an old coach and so you'd start when the mail came. And you hoped cut-out was on a mail day. There were eighteen creek crossings between here and town and there weren't any bridges, so you hoped it wasn't raining either. The road used to follow the creek bed because it was easier footing for the horses or bullock trams. The mailman used to run sly grog down here

THE BEAUTY OF THE BUSH.

Shearing in a Bar

H. P. ('DUKE') TRITTON

My shearing days are over, though I never was a gun:
I could always count my twenty at the end of every run.
I used the old Trade Union shears, and the blades were running full
As I shoved them to the knockers and I pushed away the wool.
I shore at Goorianawa and never got the sack;
From Breeza out to Comprador I always could go back;
But though I am a truthful man I find, when in a bar,
That my tally's always doubled but—I never call for tar!

Now shearing on the Western Plains, where the fleece is full of sand
And clover-burr and cork-screw grass, is the place to try your hand;
For the sheep are tough and wiry where they feed on the Mitchell grass,
And every second one of them is close to the 'cobbler' class;
And a pen chocked full of 'cobblers' is a shearer's dream of hell,
And loud and lurid are their words when they catch one on the bell.
But when we're pouring down the grog you'll hear no call for tar,
For the shearer never cuts them—when he's shearing in a bar!

At Louth I got the bell-sheep, a wrinkly tough-woolled brute,
Who never stopped his kicking till I tossed him down the 'chute.
Though my wrist was aching badly, I fought him all the way:
I couldn't afford to miss a blow–I must earn my pound a day;
So when I took a strip of skin, I would hide it with my knee
Gently turn the sheep around so the right bower couldn't see,
Then try to catch the rousy's eye, and softly whisper, 'Tar',
But it never seems to happen—when I'm shearing in a bar!

I shore away the belly-wool, and trimmed the crutch and hocks
Then opened up along the neck, while the rousy swept the locks.
Then smartly swung the sheep around, and dumped him on his rear–
Two blows to chip away the wig— (I also took an ear!)
Then down around the shoulder and the blades were opened wide,
As I drove them on the long blow and down the whipping side;
And when I tossed him down the 'chute he was nearly black with tar,
But it never seems to happen—when I'm shearing in a bar!

Now when the season's ended and my grandsons all come back
In their Vanguards and their Holdens—I was always 'on the track'—
They come and take me into town to fill me up with beer,
And I sit on a corner-stool and listen to them shear:
There's not a bit of difference! It must make the angels weep,
To hear a mob of shearers in a bar-room shearing sheep;
The sheep go rattling down the race and there's never a call for tar,
For they still don't seem to cut them—when they're shearing in the bar!

Then memories come crowding and they roll away the years,
And my hands begin to tighten and they seem to feel the shears:
I want to tell them of the sheds, of sheds where I have shorn,
Full fifty years, or maybe more, before the boys were born.
I want to speak of Yarragreen, Dunlop or Wingadee,
But the beer has started working and I find I cannot see.
So I'd better not start shearing—I'd be bound to call for tar;
Then be treated like a blackleg—when I'm shearing in a bar!

years ago. Then he'd stay the night at the huts and get on the booze and cause some terrible trouble. He always wanted to fight. He wanted to black market the grog around here but he wasn't successful. You see, I'd bring in thirty or forty dozen at the start of a shed and give the chaps a couple of bottles a night. Then we never had any trouble.

A Shearer

As far as shearers are concerned, Ken was the biggest drinker I ever knew. He drank fourteen or fifteen pints every night, before and after tea, and I never saw him drunk. For a bet, sitting in the pub, he'd drink a dozen big bottles of beer.

Ivan

Gunsmoke, the publican, he was mad. His wife went away for a few days, on the train. This insurance salesman, a Pommy bloke, went away at the same time. Now there was nothing whatsoever going on between them. But one of the shearers mentioned in the pub that both of them had gone away at the same time. This was just to stir old Cyril up and he added two plus two together over the weekend. Monday came round and old Cyril was out there on the woodheap at the hotel, waiting with a shotgun. Some of the boys from the shearing team went over to the train station and told the Pommy bloke not to come round. 'He's gunning for you.' And that's how the publican got the name Gunsmoke.

A Shearer

He'd had a couple of drinks this night and he saw a kangaroo on the road. He said to himself, 'I'll get this bastard'. Oh, he got it all right. The roo went underneath the vehicle, locking up the steering, so off the road he went, straight into a big tree He never did that again.

Ivan

We were just out of Tilpa one night, that's 120 miles (about 190 km) below Bourke, and I had the old ute with six blokes in it. When I got there and discovered I only had five I thought, 'I've lost one of these bastards on the road'. He must have stood up to have a leak and fallen off the back. He could have been 100 miles back but I found him alongside the road about 50 miles away. I'll tell you, I could have kissed him when I found him.

Another time, we were out at this shed near Tilpa—in those days the population of Tilpa was seven. There were six people in town and a bush nurse. We used to deal with this company in Bourke and when I put in the order for vegetables I said, 'We want six lettuce and make sure they are fresh'. When the grocery order arrived and came to be unpacked, there wasn't any fresh lettuce at all. Instead, they had sent six French letters. Now figure that out.

Anyhow, the shearers voted the sheep wet and since it was only two miles (about 3 km) from the shed all of them went off to the pub. I was browned off and so I went to the pub too. They had an illegal poker machine which we were playing. I could see all my money going into this machine, so as the night wore on, I took a violent dislike to it. About midnight I had another drink and said, 'I think I ought to throw this poker machine in the river.' So I got the Holden and backed it into the bloody pub door, knocking out the verandah post at the same time. The verandah sagged down where the post had been. I went inside to get the poker machine and put it in the back of the ute. The publican's wife said, 'You can't take that poker machine away, it's our livelihood'. I said, 'My bloody livelihood is already in it'. Away I went with it, out to the shed and camped there that night.

In the morning I got up, a bit sick and sorry, thinking 'I suppose I'd better take it back to these bastards'.

So I got back at lunchtime and they said, 'Where's our poker machine?'

'I got it here. What do I get in return for it?'

'Anything you like', she said, 'I'll tell you what I've got here—a plains turkey'.

You know those big wild birds they shoot? Well, she gave me the plains turkey for the bloody poker machine. If you did that sort of thing nowadays, they'd bring the bloody police in.

Norm

We'll call him Norm. He was a prudent shearer and now he is enjoying his retirement as a modest landholder. Wishing to remain anonymous, Norm was quite interesting over a couple of beers.

Individually, shearers are good blokes but collectively they are a queer kettle of fish. They don't act rationally. Anyone who earns his living with his head below his rectum has a bit of a problem. They're definitely on a handicap. You've only got a limited amount of blood in your body and when your head is hanging below your arse all day long it's at a disadvantage.

I was in a hurry to get to the shed, it was a windey old road and I'm in an old Holden ute late at night. I came round this bloody corner and I've got Buckley's of taking the bend. Going too fast. The road was up just a bit higher than the old split rail fence down below. I went straight off the road, cleared the fence, landed in the paddock on all four wheels, turned around, opened the gate and drove out onto the road. I can tell you, I didn't do the suspension much good on that old Holden. A four-foot jump in a ute! They do that sort of thing out at the Sydney Showground now. 'Hell', I thought to myself, 'I've done that on the road'.

It was so dry out west where this kid lived that in

FOUR SHEARERS AT THE END OF A RUN.

BRIAN BENSON: NO INFORMATION AVAILABLE.

JOHN MICKAN: BORN TUMBY BAY, S.A. TWENTY YEARS OF SHEARING AFTER STARTING BY BARROWING.

BILL DAVIS: ST GEORGE, QLD. INTERMITTENT SHEARING SINCE AGE FIFTEEN. ALSO AN OPAL MINER, HORSE-BREAKER AND SHARE COCKIE.

DICK LAMONT: INTERMITTENT SHEARING SINCE BARROWING START AT AGE SIXTEEN. BEST DAY'S TOTAL: 260 SHORN.

his ten years he'd never seen rain. One day he fainted and they had to throw a bucket of dust on him to bring him around.

IVAN

This particular night we went into town and stayed at the Travellers' Arms Hotel. There was a ten quid prize for anyone getting a night's sleep. You couldn't sleep there because they drank and carried on all night long. So, if you were able to sleep through the night, they gave you a ten pound prize.

Old Jim got up in the morning and asked, 'What time is breakfast on?'

This women replied, 'I'm the cook here, but the groom hasn't lit the fire'.

The groom was down back, so Jim went to wake him up. Eventually the woman put on bacon and eggs and she came in with one fried egg, one piece of bacon and one piece of bread for each of us. Apart from waiting two hours for this breakfast, her presentation was terrible. She just seemed to sling the

plates around the table at the five of us. Bob Dawson said, 'What's trumps?'

We went into Walgett and got on the grog a bit. Coming home in the truck, he just opened the door on the passenger side, got out and walked around the back of the truck. That's no great feat, so I paid no attention to him. Then he came around my side and poked his head in the window and went 'Boo'. With that he climbed down onto the running board and skinned over the mudguard, swung on the bull-bar and walked across the front of the truck, and I'm still driving fifty or sixty miles an hour (about 90 km/h). Then he spun back around and went 'Boo' into the passenger window. I'd had enough turps in me so that I didn't care whether he fell off or not. He actually stayed up there walking around for about five minutes, the mad bastard.

TYPICAL COUNTRY HOTELS, ONE NAMED TO SUPPORT THE LOCAL INDUSTRY.

12

SHEARER'S *Best* FRIEND

The working sheepdog is vitally important to shearers in particular and the wool industry in general. Just how important is the sheepdog can be illustrated by a brief story. A visitor at a shearing shed was asked to help out by filling the shearers' pens with fresh, unshorn sheep. Because the dogs excite and intimidate the already nervous sheep with their continued yapping the chap decided to attempt the job on his Pat Malone. It was pleasantly quiet as he went about his task, funnelling sheep from larger holding pens into small 'ready' pens. After the pens had been filled in what he thought was a fairly sprightly manner he was told that his efforts were not much above feeble. The advice he received: 'Next time take a dog with you. Going after those sheep without a dog is like going to a whorehouse without your old fellow.'

There is an almost reverential attitude to many man-dog relationships in the country. Canines are held in such high esteem because their labour is so valuable to the wool production team. Opinions range from one dog doing the work of three men, to one dog doing the work of an even dozen men in sheep handling duties. One could almost imagine two men and ten or twelve dogs running a 20 000 hectare sheep property. The dogs could look after the sheep on their own, the two men being necessary only to organise tucker for the dogs. It seems that every shearer has one or more dogs and exactly why these itinerant workers all need dogs is elusive. Ivan always has between two and five dogs around, at home or on the job. At the shearing shed there is a collection of dogs. Some are lying out of the way near the boards during working hours while others are tied up around the shed and at the shearers' huts.

The crown jewel of sheepdogs is the kelpie, of which there are approximately 200 000 actively working in Australia. Since there is disagreement over the origin of this breed it is appropriate to review their development.

Long-haired British dogs served the pastoralist in the early portion of the nineteenth century. They had been bred to cope with European winters, not Australian summers. Besides, they were equipped with neither speed nor endurance, traits which were necessary under the prevailing conditions. By the 1860s, shepherds were largely redundant due to the widespread introduction of fencing. The shepherds had been calming companions to their flocks. Untended, the sheep lost whatever degree of domestication they had attained in the company of shepherds. In a word, they became wild. Since they were in enormous paddocks the sheep required resolute dogs with speed and tenacity. (Dare it be said they needed 'dogged determination'?) Hence, the necessity for a new breed of working dog.

Around 1870, two black and tan collies were imported from Scotland. At about the same time another pair of Scottish collies in Victoria had a litter of pups, one named Kelpie by her ultimate owner, J. D. Gleeson. From the above dogs, and other similar pairings, arose the kelpie breed. One of these early progeny was named Barb and for a while a line of kelpies were called 'barbs'. Nevertheless, there was no difference at all between these dogs. The authority for this claim is A. D. Parsons and his words explain the breed's origin:

> As colour and type became set, kelpies were distinguished from collies, and thus came to be regarded as a separate breed. But, make no mistake, the kelpie was developed from purebred working collies. The kelpie is really a collie bred for special features now imprinted on the breed as a whole. This does not sit well with some people who equate collie with border collie.
>
> It is my firm belief that all of the characteristics of the original dogs must be credited to imported working collies—and let's not call those dogs border collies because they came later.

Parsons unequivocally states that the kelpie developed not because of breeders' efforts but 'solely by the selection of smooth-haired, prick-eared dogs of great natural ability'. One more quote from Parsons shall serve as the final word in this discussion of canine derivation :

Ever since the first day of the kelpie's existence, various writers have sought to attribute its origin to some other distinct breed of dog. There is no doubt in my mind that the kelpie did not develop spontaneously and its attributes cannot be linked or associated with any degree of certainty to anything except the working collies of Great Britain.

There are other topics of contention in regard to the kelpie. It has been said that the varying colours of the dogs means that their descent was from other breeds and other animals. Not so. The black and tans, reds, blues and creams are all a part of the collie background. Stories began about early crosses with foxes resulting in some dogs having a red colouring. It is not unusual for Scottish sheepdogs to be either red or sandy coloured.

Tomahawkin' Fred

(THE LADIES' MAN)

Now some shearing I have done, and some prizes I have won
Through by knuckling down so close on the skin,
But I'd rather tomahawk every day and shear a flock
For that's the only way I make some tin.

Chorus:
For I'm just about to cut out for the Darling,
To turn a hundred out I know the plan.
Give me sufficient cash and you'll see me make a splash.
I'm Tomahawking Fred, the ladies' man.

Put me on the shearing floor, and it's there I'm game to bet
That I'd give to any ringer ten sheep start.
When on the whipping side, well, away from them I slide,
Just like a bullet or a dart.

Of me you might have read, for I'm Tomahawking Fred,
My shearing laurels famous near and far.
I'm the don of Riverine, 'midst the shearers cut a shine,
And our tar-boys say I never call for tar!

Wire in and go ahead, then, for I'm Tomahawking Fred,
In a shearing shed, my lads, I cut a shine.
There are Roberts and Jack Gunn, shearing laurels they have won,
But my tally's never under ninety-nine.

Accounts credit the dingo with being a significant forebear of the kelpie. There is a distinct similarity in the head and ears of both animals and a certain resemblance in the lope of the two. Yes, some individuals bred together kelpies and dingoes in days gone by. But the consensus is that in almost all instances the resulting dogs were less successful than those without dingo blood. There would be no more than the merest minute trace of dingo kinship in today's kelpie.

What about the contention that kelpies contain a solid infusion of German shepherd blood? Considering the reputation of German shepherds in the Australian bush, it is hard to imagine any person being foolhardy or indiscreet enough to carry out such a blending.

SMILEY, HELPING TO FORCE SHEEP INTO DRENCHING RACE.

Did kelpies come from border collies? The answer is a resounding no. Border collies developed later, gaining a degree of fashion after the inception in Wales in 1893 of periodic sheepdog trials. That's not to deny that later breeders didn't produce crosses with border collies. Some observers feel that today's kelpies are not as outstanding as they were earlier in the century and one of the reasons for this includes the infusion of border collie blood.

Kelpie means 'water sprite' in Gaelic; it was commonly used by Scottish shepherds as a name for sheepdog bitches. Perhaps that is where the original name came from and the reason Gleeson called his dog Kelpie. These intelligent animals function admirably in what has been described as a state of 'Spartan existence'. They can be taught to respond to hand, whistle or voice signals. Kelpies can be started into training for their shepherding tasks at the age of seven to eight months. Four to six months' training turns them into first-rate sheep handlers providing, of course, that they have the raw materials to begin with. Kelpies live in the yard where a kennel gives them warmth or protection from the sun, depending on the season. Nights, they are usually kept chained. They are such keen performers that this step is necessary to keep them from going off and working the sheep on their own.

Just how resourceful the kelpie can be is enunciated by Monty Hamilton-Wilkes:

> Should a large flock approach water after a long day's trek, when there is danger of thirsty sheep trampling down the front ones in the rush to drink, the kelpies will speed once more to the front and one of the dogs will cut off the first few hundred sheep from the rest, take them to the water hole and, when they have had a drink, will move them on to graze, repeating the process with the next few hundred, while the other dogs hold the remainder steady.

Sheepdogs possess an indefatigable nature and an ever-ready keenness to get on with the job at hand. They are able to split off a portion of a mob of sheep and place them where required. They can drive sheep

MEN AND DOGS PREPARING TO PEN A MOB OF SHEEP.

'SETTING THE SHEEP', WHEREBY A DOG HOLDS SHEEP IN POSITION, UNMOVING.

into races, packing them together through side railings or by going over the top of enclosures. A good sheepdog is often considered the best asset a man can own. Their intelligence, speed and willingness is enhanced by the fact that they enjoy their work so much. Keith Willey relates a stockman's paean to a top dog:

> When you were mustering you could give a good dog several hundred sheep, an hour's walk from the bore or dam where you were assembling them, and then just go about your business and look for more. That dog would be waiting for you at the bore when you arrived and he'd have every one of his flock there and maybe some others he'd collected on the way.

To round out this brief review of Australia's own working dog, a story. Unfortunately, this is not a yarn but a true story which occurred only a few years ago. It is as cruel and uncompromising as country life can be, and as regrettable as unchecked horseplay can become. The shearer who gave this account wishes to remain anonymous; just why will soon be apparent.

Late in the afternoon a group of shearers were enjoying a smoko break when one of the station hands rode up on his horse accompanied, naturally enough, by his faithful dog. There was nothing more complicated on his mind than having a bit of a mag. As the men chatted, various members of the shearing team surreptitiously pitched pebbles at the horse's hindquarters. Not surprisingly, this caused the horse to shy and occasionally give a little kick. The station hand had no idea why his horse was a bit jumpy for he didn't notice the stone throwers. The shearers offered up a running commentary.

'Your dog's having a go at the horse's heels, mate.'

'That dog is snapping at your horse, that's why he's so jumpy.'

'Jeez, mate, I wouldn't have a dog that worried my horse that way.'

'I tell you what cobber, I'd get rid of a dog that was that much of a bother to my horse.'

This line of banter continued for a few minutes and shortly afterwards the station hand rode off.

He returned the next afternoon, on his horse but without his dog. One of the shearers asked, 'Where's your dog, mate?'

The station hand replied, 'I got to thinking about what you fellows said yesterday and you're right. A man can't have a dog that worries his horse like that. So when I rode over past the tip last night, I shot the dog.'

Kelly and Biddy

The following was penned by Keith Willey, a fine Australian writer who left us far too early in life:

> Wherever drovers and stockmen gather in their inland pubs and watering holes generally the talk will turn sooner or later to dogs, horses and women—in roughly that order. Among such meeting places was Lucy Dalton's hotel in Charleville, western Queensland.
>
> Present there one afternoon was an old man named Kelly. He belonged to that humble breed of 'smokestack drovers', so called because it was said they never travelled far enough away from home to be out of sight of their own smokestack. They might poke about between one town and another with little mobs of cattle or flocks of sheep, accompanied by a cart and two or three yapping dogs. In western Queensland towns they rated very low on the pecking order.
>
> Kelly might stand on the outskirts of a group of the 'gun' long distance drovers, listening to their conversation, but he would risk a rebuff if ever he dared put in a word. This day, as usual, the talk centred around dogs and, for just about the first time anyone could remember, Kelly mustered his courage and, during a lull in the conversation, he butted in.
>
> 'You fellers,' he said, 'Youse wouldn't ever know what a good dog is!'
>
> For a few seconds, as the saying goes, you could have heard a pin drop. Then someone spun around. 'Damn it, Kelly, you never had a decent dog in your life! What would you know?'
>
> 'Oh yes, I have had a good dog,' Kelly said. 'I used to have a bitch called Biddy.'
>
> Everybody laughed because in those parts Biddy was the name given to a legendary dog of wonderful attainments but which perhaps had never really existed.
>
> 'Tell us about her,' a drover said.

MAN, HORSE AND DOG BRING IN SHEEP FROM THE FAR PADDOCK.

'Well,' said Kelly, 'I was coming in from Emu Tanks on the Eulo road with a mob of sheep. I had a yeller-feller, a half-caste bloke, with me and when we counted them out of the brake one morning we were a hundred short.

'I said to him, "Get on your horse and go back and look for them." But as I did I noticed old Biddy heading back along the track, so I said to him, "Never mind. We'll move on. She'll find 'em."

'That night when we camped there was no sign of Biddy an' it was the same the next night and the one after that. After four days I was getting a bit worried. I thought she might have taken a bait or been caught in a trap.

'I decided I'd leave the yeller-feller with the sheep and go back and look for her. But just as I was about to saddle up, he said, "Hang on, boss. I think this is her coming."

'Sure enough, it was Biddy, and she had the sheep with her. I counted them and there was ninety-nine.'

'Hang on, Kelly,' said one of the drovers. 'You said you lost a hundred.'

'Yes, I know I lost a hundred,' he replied. 'But you've got to remember Biddy was away five days and she had a litter of pups while she was gone. So she had to kill one of the bloody sheep for rations and she dragged the pups home on the skin. That was what took her so long.'

So they shouted for Kelly in the Charleville pub, the first and probably the last time he was ever included in the 'big blokes' conversation.

TYPICAL STOCKMAN, HIS MOTORBIKE AND RED KELPIE.

PREPARING TO PEN A MOB OF SHEEP.

13

Having a MAG

It's a lovely story. It is like many of the best short stories in that it deals with simple subject matter; there are no earth-shattering issues or complicated plots involved. It is a little story containing within its folds solicitude, anxiety, sufficient minutiae to round it out nicely, suspense, death and mourning and, at the grand denouement, a thunderclap of humour which is totally unforeseen, multiplying its effect several-fold. The story has been told numerous times and each time it is ever so slightly different in its course. More than anything else it is remindful of *A Visit from St Nicholas* (*A Christmas Carol*) in its familiar, enduring nature. Tears sometimes form in the eyes of women hearing the story. It is a story that Ivan Letchford tells. It can only be described as his story. And now thoroughly primed for this wee classic, you will be left with unfulfilled expectations. The story is completely visual in orientation and would be as entertaining as the governor-general's engagement calendar if put in print. Why go to this

GETTING READY TO MOVE A MOB OF SHORN AND DIPPED MERINO WETHERS. ON THEIR RUMPS THEY CARRY THE STATION BRAND; IT IS A LANOLIN-BASED SCOURABLE BRANDING FLUID WHICH DOES NOT DISCOLOUR THE WOOL FIBRE.

BRANDING WAS ONCE COMPULSORY IN NEW SOUTH WALES BUT NOW IT IS USED IN VARIOUS COLOURS TO VISIBLY IDENTIFY THE AGE OF SHEEP IN DIFFERENT MOBS.

A RECORD LOAD OF 150 BALES BEING PULLED BY A LARGE TEAM OF HORSES. *NATIONAL LIBRARY OF AUSTRALIA.*

extent in providing background material? To strengthen a claim for ranking Ivan as a storyteller *par excellence*. Not only is he a fine yarn spinner, he is a first-rate conversationalist. When one considers the hackneyed, banal little chats engaged in each week, conversing with an entertaining, educational talker is a pleasure. That's not to say Ivan hasn't touched the blarney stone. Sometimes it's in his back pocket. When all's said and done it has been Ivan—and the Australian bush—who is responsible for this book being written. What will be done in this chapter is to bring you yarns, anecdotes, narration and opinions that have come mainly from Ivan but also from others connected with the shearing fraternity. The criterion for inclusion has been material which is both enjoyable and readable. Where Australian bush humour shines through it's a bonus, the consummate dividend for persevering readers.

IVAN

He was a really big fellow, an ill-made sort of a chap, but he could shear like hell for an hour or so. He loved tough hard shearing and he did it with brute strength.

We went to his daughter's wedding in the 1950s. He got up to make a speech and said: 'My name is Gordon McWilliams and I am a shearer and a good one at that. I wish the bride and groom every happiness. They've got £364 4s. 6d. in the bank, he's got six boxes of bees and another half-dozen nests in the bush. I wish them every happiness.' And he sat down.

When his son got married the lads at the buck's turn gave him one of those presents that is encased in a series of ever smaller boxes, the final one being a matchbox. Now this was in the days before decimal currency and you might recall that the threepence was called a 'tray'. Well, when he arrived home he opened the matchbox and inside there's a threepenny bit, a French letter and tiny, elegant card. The message read: *A silver tray and a meat cover for your wedding day.*

He married a decent sort of a girl from Werris Creek, at the Catholic church. That was on a Saturday afternoon and of course, confession was on. All the people were coming along to confession, and Gordon was out front shaking hands with the lot of them. He thought they were going to the wedding!

They reckon he shook hands with a couple of hundred people that day.

The bush was full of guys on the run from the law. In those days every time a policeman walked into the shed a bloke would disappear down the chute. Of course, they had to maintain their wives then; the women would take out a court order against them. Now they get the deserted wives pension and the blokes haven't got to disappear. There was an army of people in the bush, shearers and shedhands, that were 'wife starvers'. The police locked you up if you owed a hundred quid and you stayed there until you cut it out.

I remember I had a bloke working in the shed and a policeman came along and said to me, 'I've got to arrest this fellow.' It was for fraud or embezzlement if my memory is correct. 'You can't do that', I said to the copper, 'I'm short handed'. He said, 'I'll go back to Scone and see if I can find you a man'. He rang me the next day and said, 'I can't find anyone to replace that cove for you, I'll have to arrest him'. I said, 'Well let it go until Thursday, come later Thursday evening and I'll battle through Friday without him'. The policeman arrested him on Thursday night but that gave me a week's work out of the bloke. You certainly couldn't do that with coppers nowadays.

Oh, he was a fantastic bloke. I loved him. He'd rather have a fight then a feed. He wouldn't let anyone job me because he didn't want me to get knocked about. He was a great mate.

Anyways, in the off-season he had this old boiler and he used to whiz her over into the showgrounds at Liverpool. He told me, 'I'm in this horse stall with the bird and the caretaker whipped in with his ute, slapped it into reverse and bloody near ran over us. I jumped up and this bloke was highly incensed.

' "What are you doing there?" he asked, "I could have killed you." '

Jim said, 'Don't you worry about that. You ran over my Panama hat, you bastard!'

A Shearer

It's not the bee's knees, but then again when you've been in the game for so long if you're out of it for six months the first thing you want to do is come back to

INLAND WOOL PORTS WERE OF GREAT IMPORTANCE IN THE 1860S. BALES CONTINUED TO BE SHIPPED ON THE DARLING RIVER UNTIL THE 1930S. *NATIONAL LIBRARY OF AUSTRALIA.*

THE NORTHERN END OF THE NEWELL HIGHWAY, NEAR THE N.S.W.-QUEENSLAND BORDER.

it. If I have a holiday with three weeks off, I'm looking around for a pen. I don't know why. It just gets you. There's something about the life. It's a bug. I wouldn't miss it for quids.

IVAN

Shearers often owe you money. One time Alan gave me a hundred pairs of socks and said, 'We'll call that square, boss.' Another time he and Cliffy owed me forty quid. He owned a panel van and always had a dustcoat. One Saturday they went to Rose Hill to the races and the following Monday they had the money to pay me back in rolls of threepences and sixpences. After persistent questioning of one man as to where they got the money ... well, it's a bit of a story. Alan went around to all the food bars and said to the employees, 'I'm Mister Whoever and I want the money counted at the end of the second last race', which the people working in the stalls did. Then he just went around collecting all the money, ahead of the bloke that owned all the stalls who came round after the last race. Oh, he was a dreadful bastard.

In those bigger sheds which we don't have nowadays, I would have a bookkeeper. I'd have him weigh and brand the bales which the presser would be too busy to do, as well as looking after the bookkeeping. Not only did we have the books to do but we used to cart the number 2 stores about. That was tobacco, toilet soaps, razor blades, toothbrushes and toothpaste and that sort of gear. As a contractor I had to carry all those sorts of things around and we'd have a 'store' every Thursday night.

When we were going to a shed I'd ring the storekeeper in town and he'd deliver all the groceries. He'd also bring out all the things like razor blades, soaps and washing powders. No-one had motor cars to go to town then. Or, as an alternative, they'd come out on the mail. So then I'd have a store night on Thursday evening, which was cheque time. They got

GATES AND GRIDS DOT RURAL AUSTRALIA. THIS GRID STRADDLES AN UNDISTINGUISHED ROAD IN THE BARRINGTON TOPS AREA OF NEW SOUTH WALES.

THIS MODERN TRUCK CARRIES FIFTY-ONE BALES OF WOOL, THE INADEQUACY OF THE LOCAL ROAD PREVENTING A SEMI-TRAILER FROM HAULING AWAY 100 OR MORE BALES.

no money the first week and then 75 per cent of their money each week thereafter. On store night, they could get whatever tobacco and sundries they required for the following week. So, if I had a big group of men I'd say to one bloke, 'Okay, you can do the weighing and branding and the bookkeeping, and maybe even some penning-up.'

Ninety-nine point nine per cent of the cases in the bush revolve around your honour. If there's a drought and a chap agrees to sell you his sheep for a dollar a head and it rains tonight, he'll give you delivery of those sheep tomorrow, as promised, and there's been no contract signed. Your word is your bond in the bush and it always has been.

Dennis Ryan

There's nothing better than a hungry contractor to get work done. You walk up to Ivan the first thing in the morning and say, 'Hang on a minute, Ivan, I want to have a talk with you'. He'll say, 'Don't talk to me now, talk to me tonight'. If you get out to the shed and the cocky is there and says, 'Did you see the news last night?' Instead of starting at seven-thirty, you don't start until quarter-to-eight. When you start at quarter-to-eight that's buggered the run so there's nothing to aim for. You just poke along.

Suppose the girls from the homestead come down and start talking to you: 'Did you hear the news about the wool price?' Another bloke is whining about his hangover. Under a competitive situation you've lost it in the first half-hour. When it becomes like that, shearing becomes absolute torture. Shearing is torture but the only way shearing becomes anyways pleasant is when it becomes like the Olympic games: it's all a competition. And then when you do come to town and you are any good at all, there's old men who come and buy you a beer. Oh yes, there is a lot of fame attached to it.

A Shearer

It's changed a bit now. A few years ago, even though we were the best of mates, when we got on the board we were enemies. You'd give a bloke the hip going into the pen to beat him to a sheep. You'd run into the pen and throw a sheep out and catch it on the board. You had to have a sheep on the board when the boss blew the whistle for smoko so you'd get one more. You still find this when you get in areas where the young fellows are coming on.

We've had our days, absolutely. We're what you would call stuffed. We're still making a decent quid but not like them young fellers. They'll walk over the top of you in the big sheds. The classy young blokes will do 230 or 240 in a day. We're only old snaggers now. The young fellows learned on wide gear. We learned on narrow gear. It makes a big difference.

The Shearer's Nightmare

Old Bill the shearer had been phoned to catch the train next day
He had a job at Mungindi, an early start for May,
So he packed his port and rolled his swag and hurried off to bed,
But sleep he couldn't steal a wink to soothe his aching head.

He heard the missus snoring hard, he heard the ticking clock,
He heard the midnight train blow in, he heard the crowin' cock.
At last Bill in a stupor lay, a-dreaming now was he
Of sheep, and pens, and belly wool he shore in number three.

He grabbed the missus in his sleep and shore her like a ewe.
The first performance soon was done as up the neck he flew,
And then he turned to long-blow her, down the whipping side he tore
With his mighty knee upon her and his grip around her jaw.

And then he rolled her over, like a demon now he shore
She dare not kick or struggle, she had seen him shear before.
He was leading Jack the Ringer, he was matching Mick the Brute
When he called for 'tar' and he dumped her like a hogget down the chute.

Then he reached to stop the shear machine, all excited and out of gear
And the electric light was shining, and all was bright and clear
He gazed now out the window, half-awakened from his sleep
And down there on the footpath lay his missus in a heap.

'Gawd Blimey, I've had nightmares, after boozin' up a treat
And I walked without no trousers to the pub across the street
But this one here takes lickin' and it's one I'll have to keep
I dare not tell the cobbers I shore the missus in me sleep!'

WOOL ARRIVES AT SYDNEY'S DARLING HARBOUR, THE GREAT MAJORITY TO BE SHIPPED OVERSEAS. *NATIONAL LIBRARY OF AUSTRALIA.*

Ivan

Bill was a funny man all right. We had finished a shed and I was on my way back to Sydney. Bill wanted a lift and so we set off together They were putting bitumen on the road from Gilgandra to Dubbo. It was 1958. I can always remember because that's when they put the tar down. Now, Bill had been in the Light Horse or some other local defence force during World War II. That was the only action he ever saw. He got about as far from home as Tamworth.

There was a group of people in an old Chev car and they had run off the road and down into the ditch. One of the blokes was an Italian and he asked me if we'd give them a pullout. Hell, I got in the car and drove it out. It was on an angle of about 55 degrees and they were frightened of it. So this one Italian chap asked me if he could get a ride into Dubbo. 'Oh, sure, mate,' I said, 'get in, she'll be right'. The bloke gets in and it's immediately noticeable that he's only got one hand.

Bill is sitting there, eyeing him off, and says, 'I've seen you somewheres before.'

'Oh, you might have seen me before,' the bloke says, 'I've got a shop in Dubbo, on the corner, in the west end.'

'No, it wasn't there,' Bill said. He's still eyeing him off and then he says, 'I think I know where it was now. I think I winged you in Libya.'

Ron Taylor

My father shore at the big Belltrees shed in 1898. There were thirty or forty shearers with blades and 120 000 sheep.

It has happened. The contractor was there, shearing. He had an arrangement with the manager. (The

BECAUSE THE WORK IS SO EXHAUSTING A SMOKO BREAK IS WELCOMED AS AN OPPORTUNITY FOR A BRIEF REST.

owners of the property were businessmen in Sydney.) The contractor arranged it with the shearers that every sheep that came through showing a bit of horn was a double-header, every sheep that came through showing a few flies on them was a double-header. Sixteen men might shear 1800 sheep per day in those days but when the tally was rung up at the end of the day, every one of the sixteen shearers might get five extra for a run which is about eighty sheep, and at four runs that is 320 sheep per day that were never shorn. When I went there we were looking for 3000 sheep that were never on the books, they were knocked off. The contractor was paid for 3000 extra sheep and he split the money with the manager.

Definitely, the sheep are bigger now. The strain has been improved and there is a better quality of feed what with superphosphates and that sort of thing When I started off we were happy with six or seven pounds [about 3 kilograms] of wool per sheep and now we are unhappy if we can't get twelve or fourteen pounds per sheep.

IVAN

It makes me laugh to hear talk about post-war migrants building up Australia. It was Ron Taylor's father and the likes of him that took up the country and cleared it with an axe. You had to have £350 in the bank and an axe and if you didn't have an axe you didn't get the block of land. They are the people who physically built Australia, with their hands. It was the men who built the railway line out to Murrurundi and Dubbo, where the lines ended in those days. It was the women who had their babies way out where the nearest doctor was 300 miles (about 500 kilometres) away. It was the settlers who took up virgin blocks of country along the Darling River, west of the Darling and along the Barwon, Macquarie and Bogan Rivers. They were the people who built Australia, not the people who came along after the roads and railway lines were all in.

THE MOON RISES OVER A DROUGHT-STRICKEN QUEENSLAND PADDOCK.

A Shearer

We were shearing out at a place called Bareen. This rouseabout was a terrible lying little bastard. Anyway, when I yelled out 'raddle' he came and raddled my head instead of the sheep's.

Ivan

As a young man I used to work at a place where we pressed wool up and sent it to Boston. Scoured wool it was. I had to press the wool at the right time of day and on the draft it said the humidity was such and such an amount that hour of the day. Remember I talked about 'atmospheric regain'? If I am exporting pure wool fibre no grease, dust, dirt or vegetable fault in it and it comes out of the dryer and is slipped straight into the bale, it's a big error. I'd be sending to Boston all pure fibre, knocking myself off by 14.5 per cent atmospheric regain from the air. So, for every hundred pounds (about 45 kilograms) of wool I sent to Boston, I'd be losing 14.5 pounds because I didn't allow for the regain factor. If I pressed up the wool on a nice humid day like today, I'd be getting 14.5 per cent-plus, depending on the humidity. If I didn't allow for the atmospheric regain to occur, on the price wool is worth today, I'd be heading for poverty.

A Shearer

By Jesus, he was a tough old bastard. He was a workaholic you see. When he was out shearing at Moree his wife died. He went up to George and said 'Me wife's died'. George asked him, 'What are you going to do about it?' He said, 'I'll just let you make the arrangements', and went right on shearing. He never even went to the funeral.

Dennis Ryan

Dennis was a third-generation shearer. Now he operates a company which supplies gear to shearers:

We basically maintain that we are dealing with friends. We don't try to sell them anything. If there's any problem whatsoever, they can have two times their money back.

When I started this business, shearers would buy two pair of trousers, take them to Queensland and make sure that they got through the entire year with those two pair of trousers. Towards the end of the year they'd be getting the old emery papers coming off the grinding wheel and getting the backing—which is good drill—and patching their trousers with it.

Now nobody bothers to patch anything and we get orders for five of everything. Once there's a hole in them their wives just throw them out. Going home for the weekend means that all the laundry is done then instead of during the week.

A Shearer's Cook

Sunday was my day off and I would often saddle up a horse to meet the boys and help bring the sheep in for the last couple of miles. After a while, there would be a few sheep that wouldn't walk and so the boys would ride their bikes fast and whack them in the tail end to make them walk. If that didn't work after a couple of attempts, they'd do 'wheel stands' on their noses. And if that didn't work, they'd hang the sheep in a tree, so that the blokes coming behind would see them and bring them in after the sheep had rested.

In South Australia, any supplies we wanted came off the train. Our stop was Malboome Siding. You would have to camp overnight and when you saw the train coming you would stand in the middle of the track while waving a lamp to signal the train driver to stop. This was the procedure for people or stores.

A Shearer

Years ago, there was a big shed out Moree way called Midkin. Anyone who rang Midkin, well, that was the talk of the state. If you rang Midkin you were a top shearer. I shore there once. There were twenty-six stands. There were 57 000 sheep for twenty-six shearers. I shore 200 there on four occasions, with the narrow gear, but there were guys doing 200 to 240 a day, every day.

Dennis Ryan

I was shearing in this shed in Charleville. The team was made up of all young goers. The sheep were terrible big, dirty wethers and we were hard put to get a hundred out each day. Then a chap named 'Slogger' McIntosh arrived from town in a taxi. He started the next day and did something like 195. We all went up to 140 after that. It's psychological, a lot of it.

When I was a lad we read proper books. Most of my peers in the sheds were the same: they were all city blokes and they were all captured by that spirit of breaking with the past. When you'd get to a shed some old eccentric would open his swag and there would be a bloody pile of books, the latest and most controversial of literature. So we were all up to date with the latest ideas and the latest thoughts. You get a bloke out of some little gully up the back of the Hunter who has no affiliation with anything like this and he's an unbelievable cultural nincompoop. I'm not trying to sound like we were better, but in point of fact, we were. Once upon a time in the community, shearers were the intellectuals. Most of my mates were very sharp indeed.

IVAN

Oh yes, I've heard statements that the early graziers raped the land. It's hard for me to imagine any grazier, going back any number of years, overstocking and flogging the guts out of the land to the extent that he'd have nothing left to graze his stock on. When you do overstock you chew out all the native, drought-resistant grasses and you end up with nothing more than a heap of burrs and vegetation with no nutritional value whatsoever to the sheep. I can't see any merit to a statement that a grazier would intentionally rape his own country.

Severe damage has come about. If the grazier is stocked up and it hasn't rained, it can occur. Droughts can creep up. We've also had man-made droughts where people have overstocked and gambled on it to rain. That's what has come about in the past. Today, with sensible management, you have people who don't overstock because they only have X amount of rain per annum. These people don't put on extra stock and so they don't eat themselves out. They allow the country to 'seed up' and the sheep can live on the burr and seed which falls on the ground. It's sort of a way to conserve feed by letting the country 'seed up'.

For a while, rabbits were a terrible scourge. The owners topped up with sheep and then when it didn't rain again, the sheep ate everything that there was to eat. Then the rabbits came in and destroyed the scrub and that was the end.

14

By FAX, *Phone and* MEETING

After the concept for this book crystallised, continuing bush trips occurred, photography commenced, libraries were scoured, shearing industry workers were sought out, and the mechanics of actually writing began. During this period, changes in the shearing industry were taking place at an accelerated pace. In this last chapter, we shall look at some of these developments as well as looking forward to what we can expect of shearing life in Australia tomorrow.

Ivan expounds on the current situation:

> Since the wide comb dispute in 1982 and the introduction of wide combs into the award, the provision for weekend shearing and the influx of overseas shearers (mainly from New Zealand), we have seen suburban shearing increase to the point where it surpasses that done by traditional teams. Hastened by bitumen roads, modern transport allows shearers to return to their homes nightly, creating a watering down of unionism. Younger shearers are less involved in unions and their history, nor could they care less about the celebrated strike of 1891. Shearing used to be 100 per cent union. Nowadays nobody gives a damn about unions. The fall-out in the past five years has been dramatic.

As with many modern developments in society, outcomes are varied and frequently unexpected. Here's some of the adjustments that have taken place in shearing. Weekend shearing is allowed if shearers and shedhands are agreeable, and there has been time lost during the week due to bad weather. Shedhands were always paid for working on wet days even though they didn't work. This anomaly has been removed and shedhands are in the same category as shearers today: if it rains, no-one gets paid. To compensate shedhands for losing their wet-day allowance they now get paid for working the entire day on cut-out day even if they only work one run. It used to be acutely frustrating for all involved to get to Friday afternoon's last run and realise that there were still twenty or thirty sheep left to shear as the clock ran out. Those few sheep had to be kept penned on the weekend and then finished off on Monday morning, preventing an expeditious set-up at the next shed. It was a union rule inviolate for decades. Not any more. Commonsense has prevailed and the team can finish off the job by agreeing to work overtime Friday, giving all involved a hassle-free weekend.

One of the contributing factors to enlightenment may have been a significant agreement negotiated in 1994 between the Australian Workers Union, the National Farmers Federation and the Shearing

MOTHER NATURE REGULARLY BESTOWS VISUAL BLESSINGS UPON SHEARING INDUSTRY WORKERS. NIGHTFALL NEAR WALGETT.

Contractors Association. This accord allowed for wage conditions to be negotiated over a three-year period with adjustments to be made annually. Another player in the shearing industry has been a body seemingly misnamed PANIC, for Pastoral Awards Negotiating Industrial Committee. As Ivan explains, 'This body has removed the graziers versus shearers head-on confrontation. There has been an end to lengthy and expensive hearings in the industrial court. These days all respondents to the award negotiate by phone, fax and meeting to discuss problems, to set wages, to avoid strife.'

The good and the bad in the new conditions?

> You never knew when there was going to be an industrial problem. There might be one half-hour of shearing to do after lunch and something would crop up and you'd be stuck with fifty to a hundred sheep to shear. Another time a dispute would arise over the mess account or the cook would be unhappy about the fuel stove. You just never knew. Gone are the days when cut-out was uncertain, gone are the days when the boss had to know every trick in the book, gone are the colourful days of the itinerant shearer. All are just a memory as the modern shearer returns to the comfort of his home each evening.
>
> Today's young people just do not realise the money that was involved here, the conditions that were won. For all those years if the mattresses were too thin or if there was a hole in the meat house or the showers were inadequate, no-one

would start shearing until these things were fixed up. It was simple: it was 100 per cent union. If there was a pastoralist who presented his sheep with 'scabby mouth' (a form of dermatitis contagious to humans) there was no man in Australia who would shear them until the farmer tidied up his act. The moment the unions eased off these other blokes came in and shore anything, and on Saturday and Sunday too.

The other side of the story is that commonsense now prevails, at least sometimes. You're 200 kilometres from your next job and you're down a black soil road and it's going to rain and you can't get out of the place for a week. These days you may work an extra hour or two on Saturday morning to get the job done.

Discussing labour patterns leads me to make another prophecy. The sheep population in Australia is very low currently and if there is a rapid upsurge in wool's popularity we'll discover that there is insufficient skilled labour to meet the demand for an expanded national clip.

Ivan is greatly concerned by the fact that wool comprises but 2 per cent of the textile fibres used, that people are getting used to synthetics, that there is no longer any ignominy in wearing synthetics and that synthetics are enjoying strong popularity. He believes steadfastly that Australia must concentrate upon producing a soft-handling, stylish, attractive fibre, a wool favouring quality rather than quantity, a superior wool in the 20-21 micron range.

Peter Small, chairman of the Wool and Rural Industries Staff Training Centre in Victoria, states the situation succinctly: 'If we don't provide what the customer wants, someone else will, such is the competitive nature of this exciting global village we live in today.'

The International Wool Secretariat has found that the general public harbours negative perceptions of wool over its supposed prickliness. (This property appears, not unexpectedly, in fabrics whose fibres have incrementally greater micron thicknesses.) Major promotional activities have been carried out in North America to enhance wool's popularity. Market response to these efforts has been negative, so much so that wool's share of the North American market has dropped a compelling 40 per cent in the past eight years. 'The need for finer and more uniform wool has been apparent for decades, and yet there appears to have been no response to this market requirement', states Dr Mark Dolling of the Victorian Institute of Animal Science in Werribee. It is obvious that improvements must be made in the mobile wool factories we know as Australian Merinos.

One of the answers to improved wool production could be found in the sheep variety being developed which yields 'elite wool'. This fibre is grown on sheep with *soft rolling skin*. Not to be confused with the 'wrinkled' Vermont sheep imported from America in the late 1800s, these Merinos are a modern development with definite advantages for international markets. Soft rolling skin (SRS) produces richly nourished, long-stapled wool which is exceedingly soft and possesses a deep crimp.

SRS Merinos vary from the usual Australian Merinos in that they have many more secondary follicles in their skin, surrounding the primary ones. These follicles are closely packed and this creates wool fibres which are both finer and more uniform in diameter. Shearers will find that these sheep have higher fleece weight, less dust in the wool and to be surprisingly amenable to normal shearing techniques.

Will shearing itself remain a principal vocation in Australia? The answer, for the foreseeable future at least, is an emphatic yes. Shearers will be aided by enhanced technology as their handpieces continue to become more efficient over the years. Comb function has likewise improved considerably in recent times; as Ivan relates, 'the design of modern combs now literally gathers in the wool for better cutting'.

Two generic alternatives to shearing wool from sheep have been placed in the too difficult category, at least for the time being. Robotic shearing never really got off the ground because of the great physical differences in sheep. Diversity in conformation, size and covering created an ill-defined model that failed to mesh with the realities of the sheep.

A program for biologically removing wool showed much early promise but was unfortunately disbanded. This methodology evolved around a halt in wool growth for a period of eighteen hours, the cessation resulting in a 'break' in the wool. In order to keep the wool from falling off the sheep at an inopportune moment, the sheep were to be fitted with a form of net wrap. Then, at the propitious time the animal would be installed in a 'doffing' machine and the wool would be vacuumed off. That was the plan, but since it has so far defied solution we are going to continue to have shearers making four runs a day, five days a week and sometimes, now, on Saturday mornings as well.

Ivan now assumes the role of soothsayer. 'My prediction is that eventually it will become every man for himself, an occurrence that will not benefit the industry overall. Even "every man for himself" has become a hollow phrase as another significant change is the presence of women in the woolshed. It seems that women now make up the majority of shedhands. Frequently the husband or the boyfriend is a shearer and the lass becomes a rouseabout. There is little reason to think that this feminisation of the shearing industry will abate in the years ahead.'

If only Henry Lawson could return from the grave and reflect upon his poem 'The Shearer's Dream". Verse by verse, portions of the dream have eventuated with the passage of time.

For every one of the rouseabouts was a girl dressed up as a boy . . .
They had flaxen hair, they had coal-black hair—and every shade between . . .
The shed was cooled with electric fans that was over every shoot . . .
The huts had springs to the mattresses, and the tucker was simply grand,
And every night by the billabong we danced to a German band.

APPENDIX

BALES OF WOOL PACKS WAITING TO BE FILLED WITH NEWLY SHORN WOOL NEAR THE DOOR OF A SHEARING SHED.

Shearing a Sheep

Roy Jerrim was until recently a shearing instructor working under the auspices of the Australian Wool Corporation. An ex-Queenslander, Roy is a resident of the Australian Capital Territory, the base from which he taught approximately 400 youths how to shear efficiently. Roy began as a rouseabout and went on to become a gun shearer with tallies of 300 per day on narrow gear. After thirty-three years of shearing and being placed in numerous competitions, Roy was involved for ten years in preparing tomorrow's professionals. Here's his blueprint for shearing a sheep.

Correct Shearing Position

The correct starting position is most important.

Step 1: Sheep and shearer in upright position. Shearer's right knee opposite downtube, and knee approximately 300 mm out, opposite elbow on downtube.

Step 2: Move backwards, right foot slightly more so than left. Laying sheep over onto its right rump, create a stretched belly, enabling the comb to travel over a smooth area of skin.

Step 3: Place sheep's front right shank high against your hamstring, knees against sheep. Both sides of brisket are visible. Keep sheep's head under control.

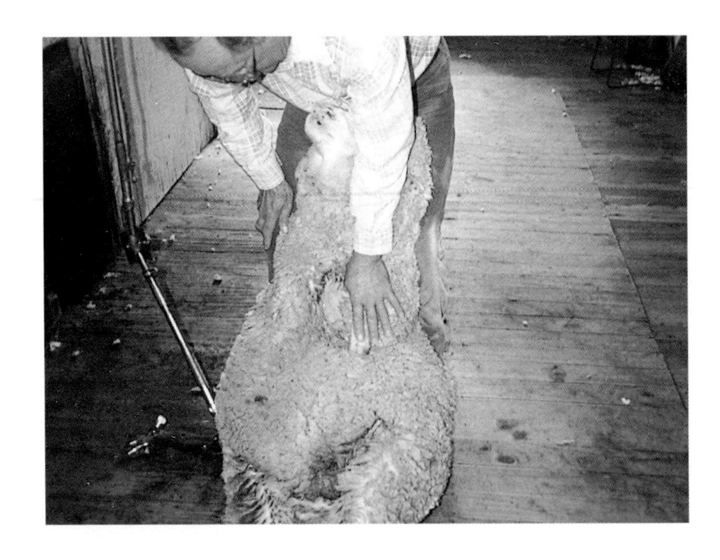

Belly Wool Removal

Remove only belly wool. Don't shear too wide out into the fleece wool on each side. As well, do not shear too narrow, which results in leaving belly wool attached to the fleece.

Crutch

Place a small outward stroke near the top of the sheep's right hind leg. Then start the crutch blow from the inside shank around crutch to inside left shank. Care should be taken around the teats.

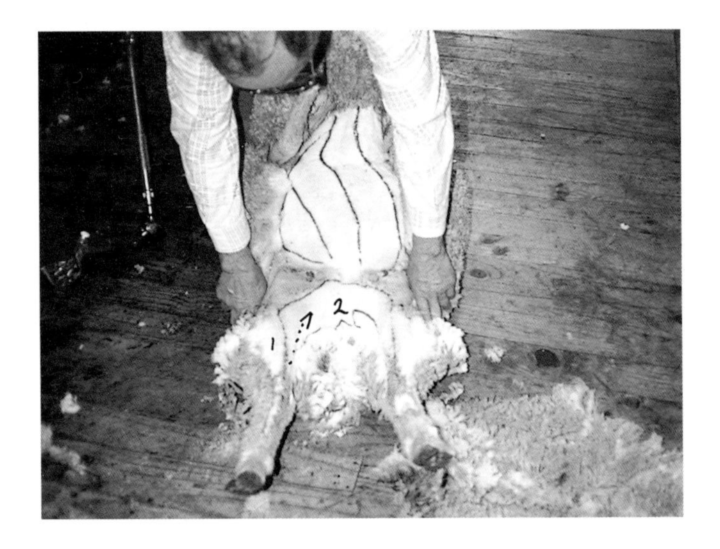

First Hind Leg

Start with a small stroke into the flank, making first blow on long blow easy. With sheep sitting upright, move backwards slightly with each following stroke until the first leg is completed.

On the first blow, aim from the flank to the floor, making sure to finish on the skin and about a comb-width short of the backbone. On the second blow, start from the shank and finish at the same depth. The same applies to the third and fourth blows. After clearing the tail, place the first blow parallel to the floor, above the backbone, finishing to match the first down stroke. The next and final blow is over the backbone; it finishes with the same length. The right foot must move well back with successive blows.

Top Knot

Clear away both eyes and top of each ear. For footwork assume the same position as in finishing the first hind leg.

Neck

Make a small clearing blow at the top of the brisket area and under the first fold (on Merinos). Continuing over the fold, keep on the skin with the bottom tooth well down The top tooth follows the throat line (not over it), continuing to under-chin and around face to the ear. Keep your left leg straight with this stroke

For the next blow start deep with a full comb, bottom tooth hard down. Continue up behind the head, clearing the ear first. Placing another small stroke well behind the ears will give the shearer more clearance for his left hand on the long blow, making it easier to bow the sheep around his left leg. The left leg is well back by this time, against the sheep's front legs and close to its brisket.

First Shoulder

This can be done in two blows if the left hand keeps the shank low, not lifting. Start the first stroke with only the bottom tooth picking up fribs at the top of the leg, then gradually fill comb to the required depth, which is level with or matching the last neck blow. The next stroke may start with a full comb but finishes at the same depth as the first one. Another small clearing blow may be necessary, moving the sheep while doing so.

Long Blow

Position is most important, for animal and man. Sheep well up on its backbone. Shearer's right foot placed in the crutch, near the tail. The left toe is under the sheep where its shoulders join and the knee is against the brisket area. Start the first blow with a full comb, continuing with a slight uphill direction to finish near the side of the brisket. Each stroke from now until the backbone is reached should be of the same width. It may be necessary to use a narrow blow at the backbone.

A lot of second cuts can occur as a result of a badly placed blow. Once again, footwork is most important. After three blows both feet must continually move backwards. Keep the left foot tightly placed against the sheep's front legs. After the fourth blow put the right foot over the sheep's rump, pulling slightly with the heel just in front of the tail, imparting a rolling motion to the animal's body.

Over Backbone

Again, a lot of footwork is necessary to finish in the correct position. Always start with a full comb, the bottom tooth down. Keep a firm grip, with the left hand holding the sheep's head behind the ears, bringing its head and neck around your left leg. If considerable pressure is felt against the left foot it may not be back far enough against the sheep's front legs.

Over Head to Last Shoulder

Clear area around face, then continue straight downwards to brisket area. With remaining blows keep the top tooth just out of the wool; the comb should take the remaining wool quickly and with ease. Don't go too deep. Try to finish on a straight line which is level with the shoulder line.

Around Last Shoulder

Only two blows are needed in this area providing you start with a full comb. The first blow goes out to the shank knuckle using the thumb to feed top foreleg wool into the comb. Clear the shank and by holding it high, perform a full blow from the shank down and around the bare skin under the foreleg, being careful not to cut the bare skin. Continue down to the flank and if the shearing is easy continue on out to the toe.

Last Quarter

The sheep must be laid back toward your right leg to successfully execute this blow. If blow is placed correctly the finish is simple. Just start with a full comb, keeping the bottom tooth down. Attempt to finish with a full comb, then finish blows under the leg, thus completing the last side very quickly and with little effort.

Roy concludes with his personal keys to good shearing:

- Maintain well-ground and properly selected combs and cutters.
- Choose straight-bottom tooth combs for merinos, wider models for crossbreeds.
- Shear sheep in the correct position.
- To avoid second cuts keep bottom tooth pressure on at all times.
- Keep top tooth just out of the wool after opening blows.
- Do limbering-up exercises before each run.
- Use correct clothing and boots.
- Keep the lower back region covered and warm.
- Avoid sagging beds.
- In hot weather drink frequent small quantities of temperate water.
- Maintain good health. Eat correctly. Don't overeat in hot weather. Ease back on food intake when not shearing.

Good shearing!

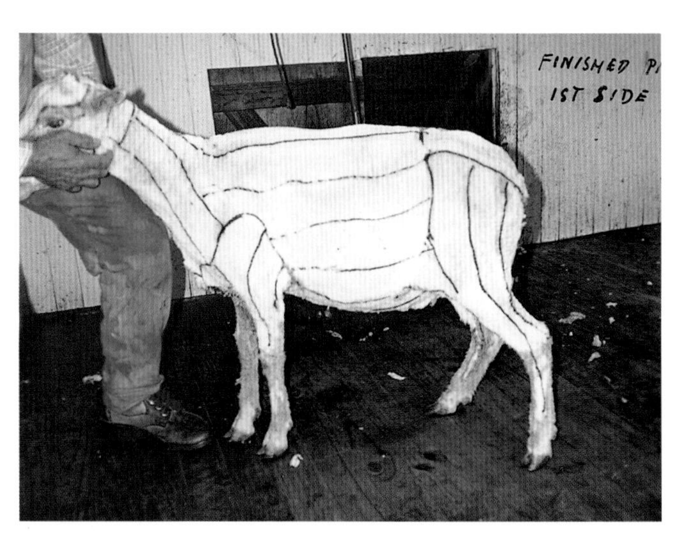

GLOSSARY

'All on the Board'
A cry executed when the final sheep of a mob are in the catching pens.

Bag Boots
Mocassin-like footwear favoured by shearers, so named because they originated from the jute woolpacks.

Barrowing
A method of learning to shear whereby the beginner starts or finishes off the sheep for a shearer.

Blades
Idiomatic term for hand shears, used before machine shears became commonplace late in the nineteenth century.

Blow
A continuous stroke of the shears along the sheep's anatomy. Most commonly used in the self-descriptive term, 'the long blow'.

Board
The floor of the woolshed, upon which the sheep are shorn.

Bogghi or Boggi
Handpiece of shearing machinery. Early versions resembled the bogghi lizard, hence the name.

Boss's Boots
Working bent over with head down, the first sight of the boss by a shearer. (To watch for the boss's boots.)

Branding
Relevant information stencilled on wool bales; identifies property, classer, contents and number.

Broomie
General hand on the board responsible for wool-sweeping tasks.

Classing
The job done by the classer whereby the shorn wool is identified and graded

Clip
Total amount of wool shorn from one place in a season. It could be the single clip from a single property or, at the other end of the scale, the national clip.

Cobbler
A sheep which is very difficult to shear, commonly the last sheep left in the pen.

Core Test
Sample taken from a bale of wool. Modern method of assessing wool on a scientific basis.

Cotted
Used to describe a fleece with deficiency wherein the wool fibres have become matted or felted.

Crutching
Removal of wool from area of sheep's crutch; lessens the danger of blowfly attack.

Cut-Out
The completion of shearing.

Dag
Wool on the rear quarter of a sheep, usually dirty with sheep dung and mud.

Double Fleeced
Used to describe sheep with two years' growth of wool.

Expert
The individual who grinds the combs and cutters and ensures that the shearing shed machinery is running smoothly.

Flock
Total of all the sheep run on a property.

Flyblown
Pertaining to sheep, it means the malady resulting from maggots of fly eggs invading the animal.
Pertaining to men it means completely penniless, dead broke.

Fribs
Stringy bits of wool which grow beneath the sheep's arms and in the crutch.

Gun
An extremely fast, competent shearer.

Jacky Howe
A sleeveless shirt favoured by many shearers. Ironically, Jack Howe never wore these eponymous versions; instead he chose shirts with snug sleeves.

Lambing Down
The name given to the process whereby unscrupulous publicans separated a shearer from his cheque. A nineteenth-century phenomenon.

Lanolin
The natural lubricant of the wool fibre; of significance to the cosmetics industry and a component of many medicaments.

Locks
Fribs, sweatlocks and second cuttings together comprise locks when shorn wool is labelled.

Merino
The breed of fine-woolled sheep originating in Spain but only attaining pre-eminence in Australia.

Mob
Those sheep of the same breed that have run together since the previous shearing.

Mulesing
A skin removal operation in the breech area of the sheep, conducted to lessen blowfly attack.

Muster
A round-up or gathering together of sheep.

Noils
Short, brittle wool fibres removed during the combing stage of textile preparation.

'Off the Dog'
Shearers won't work with freshly yarded sheep, i.e. those fresh 'off the dog'. The sheep must be penned for at least four hours, making them easier to shear.

Paddock
An area enclosed by a fence; can be small or vast.

Palliasse
A mattress of strong fabric stuffed with straw; until 1947 shearing team members slept on palliasses.

Pen
Enclosure for holding sheep in woolshed environs. There are sweating pens, holding pens, catching pens and counting-out pens. A newly hired shearer obtains a pen.

Penner-Up
The individual who fills the catching pens with sheep for the shearers.

Picker-Up
Shedhand who picks up freshly shorn fleece and throws it onto the classing table.

Pink 'em
Very close, high-quality shearing. Owners always hope the shearers pink 'em.

Pizzle
The penis of a sheep.

Pressing
After wool is classed the presser packs the wool into bales ready for transporting.

Raddle
Coloured chalk to mark sheep for various reasons. Historically, sheep were raddled by squatters to

HIGHWAY BRIDGE OVER RIVERBED TEMPORARILY REDUNDANT.

indicate sheep they felt weren't shorn properly; thus they avoided paying shearers for work done.

Rep
Short for representative; the shearers' elected or delegated nominee in any union-management discussion.

Ringer
The fastest shearer in any shearing shed.

Rouseabout
General shedhand, anything from a board boy to a woolroller; in common Australian fashion, often shortened to 'rousie'.

Run
The official time for a period of shearing, a run is an unbroken two hours There are four runs per working day.

Scour
1. To wash wool, removing impurities from the fibres.
2. A sheep with the 'scours' is experiencing diarrhoea due to an illness or perhaps a change of pasturage.

Second Cuts
Short pieces of wool which result from a shearer not cutting close enough, thus having to go over the sheep a second time.

Shearers: Suburban or Traditional
Suburban shearers travel to and from their work on a daily basis; they reside in the area in which they are employed.

Traditional shearers move about the countryside as a team, remaining on a property until the shearing is completed.

Shearing: Contract or Cost-Plus
In the contract system of payments a price is agreed upon before work commences. It involves a predetermined fee.

In the cost-plus system of payments the men are paid and then the contractor adds his margin to the cost after the sheep have been shorn. It involves an indeterminate fee.

Shedhand
See Rouseabout.

'Sheepo'
Call given by shearer when the last sheep has been removed from the catching pen; indicates pen requires immediate restocking. Also refers to a shedhand employed to keep the pens full.

Smoko
Much anticipated thirty-minute tea-break.

Snagger
A rough untalented shearer.

Snob
Last sheep in the pen; often the snob is a cobbler.

Snowed In
Describes the situation when shearers get well ahead of support staff and classer; workers are 'snowed in' with fleeces piled about, waiting to reach the wool table.

Staple
A natural cluster of wool fibres grouped together.

Sweat Extractor
Rhyming slang; sweat extractor = contractor.

Sweatlocks
Second cuts or single strands which become separated from the rest of the fleece.

Sweats
Waxy, grease fibres originating in the crutch areas.

Tally
Number of sheep shorn in a period, anywhere from a run to the duration of a shed; recorded in a tallybook.

Tomahawker
Evocative label for the rough, careless shearer.

Weaner
A sheep weaned from its mother; usually six to nine months of age.

Wet Sheep
Refers to sheep which cannot be shorn because their fleece is too damp. Shearers cast a ballot on whether sheep can be shorn (dry vote) or whether they will have a day off (wet vote).

Wigging
The cutting of wool from the area surrounding the sheep's eyes.

'Wool Away'
Call to rouseabouts from the shearer, indicating the last fleece has not been cleared away, thus interfering with the shearer's work.

Yolk
Generic term for all of the yellowish greasy matter in unprocessed wool; when heavy it tends to plug-up shearer's combs. Lanolin is a by-product of yolk.

TO GET TO REMOTE SHEDS SHEARERS MUST OFTEN DRIVE ON MEDIOCRE ROADS. SOMEONE SERIOUS ABOUT ROADWAY DEFICIENCIES IS OBVIOUSLY EXASPERATED WITH HIGHWAY OFFICIALDOM.

BIBLIOGRAPHY

Books

Adam-Smith, P., *The Shearers*, Thomas Nelson, Melbourne, 1989.

Anderson, R., *On the Sheep's Back*, Sun Books, Melbourne, 1966.

Barber, A. A. and Freeman, R. B. *Design of Shearing Sheds and Sheep Yards*, Inkata Press, Melbourne, 1986.

Bean, C. E. W., *On the Wool Track*, Angus & Robertson, Sydney, 1910.

Blainey G., *A Land Half Won*, Macmillan, Sydney, 1980.

Blainey G., *Our Side of the Country*, Methuen Haynes, Sydney, 1984.

Boldrewood, R., *Shearing in the Riverina*, 1865, Halstead Press, Sydney, 1878.

Bowen, G., *Wool Away*, Whitcombe & Tombs, Auckland, 1955.

Breckwoldt, R., *Wildlife in the Home Paddock*, Angus & Robertson, London, 1983.

Burfitt, C. T., *History of the Founding of the Wool Industry in Australia*, 1913.

Clune, F., *Search for the Golden Fleece*, Angus & Robertson, Sydney, 1965.

Curr, E. M., *Recollections of Squatting in Victoria.* George Robertson, Melbourne, 1883.

Day, G. and Jessup, J. (eds), *History of the Australian Merino*, Heinemann, Melbourne, 1984.

Dutton, G., *The Squatters*, Currey O'Neil Ross Pty Ltd Melbourne, 1985.

Ferry, J, *Walgett Before the Motor Car*, printed by Walgett Shire, Walgett, 1978.

Freeman, P., *The Woolshed: A Riverina Anthology*, Oxford University Press, London, 1980.

Garran J. C. and White, L., *Merinos, Myths and Macarthurs*, Australian National University Press, Sydney, 1985.

Guthrie, T. F., *Australia's Greatest Industry, Sheep and Wool: A Brief History*, Melbourne, 1939.

Hamilton-Wilkes, M., *Kelpie and Cattle Dog*, Angus & Robertson, Sydney, 1967.

Harrison, J., *Pulling the Wool*, Child, Henry & Plumb, Brisbane, 1987.

Hawkesworth, A., *Australasian Sheep and Wool*, William Brooks & Co., Sydney, 1920.

Hewat, T., *Golden Fleeces: The Falkiners of Boonoke*, Bay Books, Sydney, 1980.

Hewat, T., *Golden Fleeces II*, Bay Books, Sydney, 1986.

Holthouse, H., *Up Rode the Squatter*, Rigby, Adelaide, 1970.

Livestock and Grain Producer's Association of New South Wales, *The Sheep Production Guide*, New South Wales University Press, Sydney, 1976.

Luck, P., *The Australians*, Lansdowne Press, Sydney, 1981.

Magoffin, R., *Waltzing Matilda*, Australian Broadcasting Corporation, Sydney, 1983.

Merritt, T., *The Making of the Australian Worker's*

Union, Oxford University Press, London, 1986.
Palmer, J. A., *The Great Days of Wool, 1820-1900*, Rigby, Adelaide, 1980.
Parsons, A. D., *The Working Kelpie*, Thomas Nelson Australia, Sydney, 1986.
Pemberton, P. A., *Pure Merinos and Others*, Australian National University, Sydney, 1986.
Rolls, E. C., *They All Ran Wild*, Angus & Robertson, Sydney, 1969.
Ronald, H. B., *Wool Past the Winning Post*, Landvale Enterprises, Melbourne, 1978.
Sowden, H., *Australian Woolsheds*, Cassell Australia Limited, Melbourne, 1972.
Taylor, P., *Pastoral Properties of Australia*, George, Allen & Unwin Australia Pty Ltd, Sydney, 1984.
Taylor, P., *Springfield: the Story of a Sheep Station*, George, Allen & Unwin Australia Pty Ltd, Sydney, 1987.
Townend, C., *Pulling the Wool*, Hale & Iremonger, Sydney, 1985.
Tritton, D., *Time Means Tucker*, Shakespeare Head Press, Sydney, 1964.
Ward, R. and Robertson, J., *Such Was Life*, Ure Smith, Sydney, 1969.
Whitlock, R., *Rare Breeds*, Prism Press, Dorset, England, 1980.
Wignell, E., *A Bluey of Swaggies*, Edward Arnold Australia Pty Ltd., Melbourne, 1985.
Willey, K., *The Drovers*, The Macmillan Company of Australia, Melbourne, 1982.
Williams, A., *Backyard Sheep Farming*, Prism Press, London, England, 1978.

Other

Australian Wool Corporation, *Report to International Wool Textile Organisation*, June 1987, Rio de Janiero.
Hyde, Nina, *National Geographic*, Volume 173 No. 5 May 1988, pp 552-583.
National Material Handling Bureau, *Wool Handling, study of the handling and movement of wool from sheep's back to broker's store*, 1970.
National Wool Producing Industry Training Committee, *Shearer Training in the 1980s*, Melbourne NWPITC, 1979.
New South Wales Wool Producing Industry Training Committee Ltd, *Report on Shearing School conducted at Murtee Station, Wilcannia, New South Wales, August 1983*, T. Latchford, October 1983.
Pastoral Industry Award 1986, P 144 AM Print G 6783, The Australian Conciliation and Arbitration Commission.
Record of the Industrial Dispute Between Australian Worker's Union and Farmer's Union of Western Australia, Industrial Association and Tasmanian Farmers' Federation Employers Association; POOI MIS 186/84 MD F5655.

INDEX

S

T

W

Y